Electricity for All

The Story of Ozark Electric Cooperative, 1937–2012

By

Jim McCarty

The Donning Company Publishers
184 Business Park Drive, Suite 206
Virginia Beach, VA 23462

Steve Mull, General Manager
Barbara Buchanan, Office Manager
Anne Burns, Editor
Nathan Stufflebean, Graphic Designer
Priscilla Odango, Imaging Artist
Cindy Smith, Project Research Coordinator
Tonya Washam, Marketing Specialist
Pamela Engelhard, Marketing Advisor

Neil Hendricks, Project Director

Library of Congress Cataloging-in-Publication Data

McCarty, Jim, 1962-
Electricity for all : the story of Ozark Electric Cooperative, 1937-2012 / by Jim McCarty.
p. cm.
ISBN 978-1-57864-762-0
1. Ozark Electric Cooperative. 2. Electric utilities--Missouri--Lawrence County--History. 3. Electric cooperatives--Missouri--Lawrence County--History. 4. Rural electrification--Missouri--Lawrence County--History. I. Title.
HD9685.U7O963 2010
334'.681333793209778835--dc22

2012011598

Printed in the United States of America at Walsworth Publishing Company in Marceline, Missouri

Table of Contents

Mary Ann Moore Creed, left, and Sue Childress Compton were two of the many former Ozark Electric employees who contributed to this book.

Acknowledgments

This book would not be possible without the diligent assistance of Ozark Electric's Linda Stander, who left no stone unturned in her quest to find historic items to feature in this book. In addition, and fully knowing this list will leave someone out, the author wishes to thank the following people for their assistance:

Amy Bass, Empire Electric
Sue Childress Compton
Mary Ann Moore Creed
Don Eskrew
Charles Fender
John Lomas
Jerry Mayberry
Joanne McCann
Bob Neal
Pat Prewitt
DeLana Stokes
Jack Swearingen
Gene Thomas
Anne Williams, KAMO Power
Don Wood
Stephen Wright

Introduction

When the original organizers of Ozark Electric Cooperative first met in March 1937, they had an ambitious plan in mind. They wanted to electrify not just a single county as most of the earlier electric cooperatives had done. They were planning to wire the entire landscape of southwest Missouri, from as far to the east as Webster County, south to the Arkansas line, north to Camden County, and west to the Oklahoma border.

Had the Rural Electrification Administration not stepped in to limit its service area, Ozark Electric would today occupy more land area than any other system in the state. Instead, its early efforts gave birth to six other electric cooperatives. And Ozark Electric remains one of the largest and fastest growing in the state, serving more than 30,000 members over 4,763 miles of line in nine counties and four small towns with ninety employees.

The seventy-five-year story of Ozark Electric Cooperative is one of triumph over tremendous adversity. It's one of a people who wouldn't take no for an answer, shrugging off the suggestion that mere farmers couldn't organize, much less operate, such a challenging business. In the end, Ozark Electric not only proved its critics wrong, it also thrived through downturns in the economy, wars, droughts, ice storms, tornadoes, and fierce competition from its rivals.

That it has remained such a vital force in the community is a true testament to the men and women who formed it and kept it strong over the years. It's fitting that on the seventy-fifth anniversary of this cooperative, we take a look back at this great accomplishment.

As the nation struggles to bring itself out of its current financial crisis, this book is being written to highlight the efforts of those who made Ozark Electric Cooperative such a success. Perhaps the story of this business, which began as a partnership between government and the people during one of the most challenging economic periods in our nation's history, can serve as a model for those working on solutions to today's economic woes.

Ozark Electric exists for only one reason and that is to provide service to its members. Read on then and join us in recalling the triumph of human spirit that made Ozark Electric what it is today.

Jim McCarty
January 2012

This was Ozark Electric's fleet in the early 1950s.

Meals by lantern light were not romantic affairs for those living without electricity. Often families only had one lantern, forcing some to sit in the dark when the light was needed elsewhere. Photo courtesy of the Library of Congress.

Chapter 1
The Dark Days

If you live in Ozark Electric's service area today, you can count on a steady supply of affordable, reliable electricity. You can own a wide range of electric equipment, from large screen televisions, electric ranges, mixers, computers, even heated toilet seats if you want, and never worry about whether you will have the electricity to power these devices.

If you lived in the same place seventy-five years ago, things would have been much different. As late as 1937, rural people in southwest Missouri lived little different than their pioneer ancestors. Their lives were a constant drudgery from the moment they awoke—in the dark—to the time they retired early in the evening because it was too dark to see.

If you lived in rural southwest Missouri before Ozark Electric was formed, you were well acquainted with pumping and carrying water by hand. Because there was no electricity, there was no indoor plumbing. Water was stored in cisterns or dug wells and laboriously pumped into buckets, which were then carried to the barn to water livestock or into the house for cooking and cleaning.

Wash day involved more carrying of water and was usually done once a week. This was done outside so that water could be heated on wood fires. Then the clothes, grimy from farm work, had to be scrubbed on wood and metal washboards. Woe to the housewife who let her knuckles touch the metal edges! The clothes were hung on outside clotheslines, if it was warm enough, or all over the house in the wintertime.

Likewise, bath time took place once a week, often on Saturday before the family headed into the nearest town for provisions. A wooden tub served as the bathtub for many families, and bath water was often used for more than one person, giving rise to the saying, "Don't throw the baby out with the bath water." Once again, water had to be heated over a fire and carefully carried to the tub.

In dairy country, herd size was limited to the number of head that could be milked by hand. Often the milk spoiled because there was no refrigeration. "I've gone many a time to the barn with a lantern on my arm," recalled Ozark Electric pioneer Raymond Phariss, an early director who was interviewed for a video in November of 2005 when he was one hundred years old. "I remember my mom would be milking in the lean-to on the back of the barn. I would borrow the lantern to throw hay down, and she would have to milk in the dark."

Families had to be creative to keep food from spoiling. Hams were salted and cured for later use. Dairy products were lowered down a well or kept in springhouses. Smoking meats helped to preserve them, and smokehouses were common on farms.

In the days before Ozark Electric was formed, clotheslines were used to dry clothes. In the winter they were hung inside wherever space could be found. Photo courtesy of the Library of Congress.

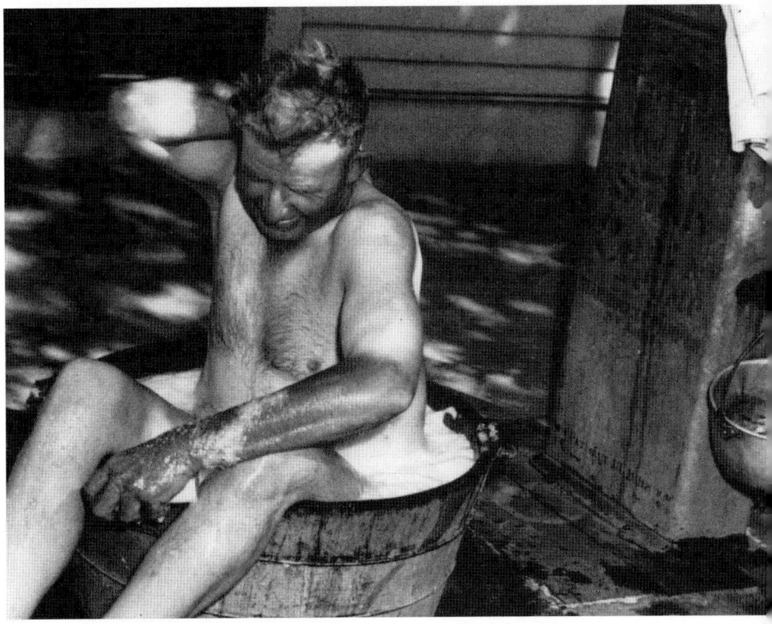

With no indoor plumbing, rural families bathed once a week or less, using a washtub and water heated on the woodstove. First in line got the cleanest water. Photo courtesy of Association of Missouri Electric Cooperatives.

Wash day was hard work for families with no electricity. Usually water was heated in a large iron pot. Clothes were scrubbed on a corrugated washboard. This, along with the strong lye soap, caused cracked, red hands for rural women. Photo courtesy of the Library of Congress.

The aptly named "sad iron" was a chunk of cast iron that was heated on the stove. Rural women often had multiple sad irons so they could switch to a hot one as the one they were using cooled. One had to be careful not to dirty clean clothes with soot from the stove. Photo courtesy of the Library of Congress.

Cooking on a woodstove was an art requiring the cook to adjust the heat using dampers or the type of wood. The stove heated the house during the summer but barely drove away the chill in the winter. Photo courtesy of the Library of Congress.

Since there was no electricity, dangerous candles and kerosene lanterns were used for lighting. Many a barn was burned down when an unruly cow kicked over the lantern. Rural kids had the task of cleaning soot off the glass chimneys on the popular Aladdin lamps.

Cooking is another task that today's homemakers could not relate to. Because there was no electricity, wood cookstoves were the only option for rural families. In the summer, these massive cast-iron appliances heated a lot more than the food. In the winter, their warmth barely spread to the table. It took a skilled cook to keep the fire going at just the right temperature to ensure even baking without burning the food. A plentiful supply of firewood had to be cut, split, stacked, and brought into the house for the cookstove as well as any heating stoves that were in use.

Outhouses were another staple of rural life in the 1930s. For obvious reasons, they were built downwind and as far from the house as possible. In the summer, farm families battled snakes and wasps for the rights to the little "necessary room." In the winter, they stuffed newspapers in the cracks and took care of business as fast as possible in the bitter cold.

Bill Mieswinkel of Mount Vernon recalled life without electricity in his submission for the book *The Day the Lights Came On*, which was published by the Missouri Rural Electric Women's Association:

Before electricity came to our house, we had kerosene lamps which smelled, the globes had to be cleaned constantly, wicks had to be trimmed (just so) and once one was broken while it was burning and we had a small fire. We never had enough lamps for five people all at one time; so we stayed in one or two rooms. We used a kerosene lantern to go to our outhouse about 50 yards from the house. Many times the wind would blow the lantern out and then you had to return to the house and relight it. The rain would crack the globe when you had to work outside at night. Just a trip to the barn could be a real problem.

Springfield's Lorena Kemppainen also recalled those "dark days" in the book:

Before the lights came on, necessity created a LOT of family togetherness as our family of six huddled around the kitchen table. Those were the days of the "Big Depression"—so when homework was finished we went off to bed to save kerosene—the only exception being the nights when the appeal of bright moonshine lured us outside.

Advertisements like this one from Western Electric Company caused rural housewives to dream of a better life with electricity. "Electricity should be your servant," this ad stated. Photo courtesy of *Rural Missouri* magazine.

There was no radio—no television. We played records on a wind-up phonograph the size of a small suitcase. Otherwise, we provided our own fun. How did we keep the milk from spoiling? It was poured into jars which were then placed in a bucket and hung in a cistern to keep cool. Water was carried into and out of the house in buckets—there was no kitchen sink or kitchen counter, and a bathroom did not exist. An outhouse (a two-holer that created a fly problem during summer months) was the only means of disposing human waste. At night we went there using a flashlight because of snakes and black widow spiders that had a tendency to build their webs across the holes.

We bathed once a week—Saturday night. Water heated by the kitchen stove reservoir was then poured into a galvanized tub placed on the floor near the stove. The same start-up water was used, but warmed up for each person with hot water poured from a tea kettle simmering on top of the stove. In the dim light of the kerosene lamp, privacy was not a problem.

With no electricity to power pumps, water was pumped by hand from cisterns and shallow wells. Children often had the job of fetching water for use in the home and for livestock. Photo courtesy of the Library of Congress.

> Clothes were washed outdoors on Monday, weather permitting. My earliest memory is of Mom and Grandma washing clothes with water that had been heated in a large iron kettle over a campfire. They used washboards for the process, which took most of the day. Because of the heavy bib overalls in the wash, their hands were often rough and swollen.
>
> During winter months, the heat from the wood stove in the kitchen did not penetrate into the back bedroom, so we were sent to bed carrying a rock or brick to place at our feet after it had been heated in the oven, then wrapped in a thick cloth.

Don Wood, former Ozark Electric assistant manager, recalled those dark days in a column he wrote in 1985: "I recall as a boy growing up in the late '40s in western Lawrence County. Mother would rise early in the morning and fix breakfast over a wood cook stove. After breakfast, the cows had to be milked by hand, water to cool the milk was pumped by hand, and water for drinking, laundry, bathing and cooking was also pumped by hand. One thing we didn't pump was water for the restroom—only because it was located in a small building at the end of a path."

Life was hard in those days. With only 10 percent of rural homes wired for electricity, compared to 80 percent of urban homes, all chores had to be done by hand. There had to be a better way. Rural families looked for the light and saw it on the horizon, where the glow from cities and small towns beckoned them to a better life. Small wonder rural kids left the land for jobs in the cities whenever they could.

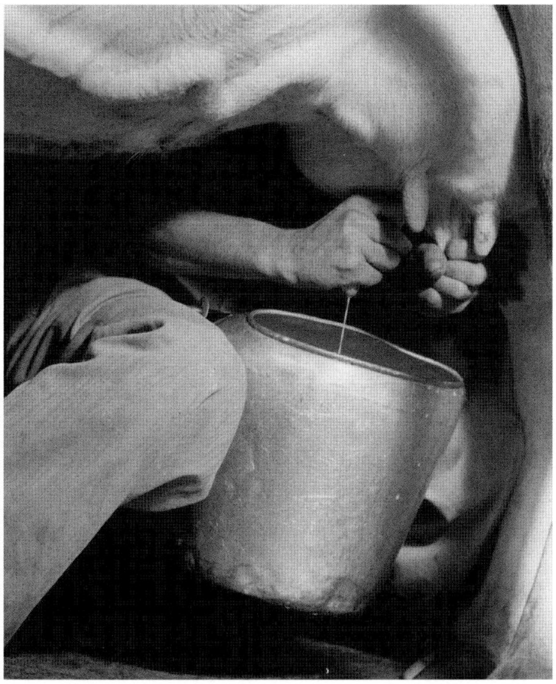

The strongest voices calling for electricity for the farm came from dairy farmers, who struggled to milk cows by hand. Photo courtesy of the Library of Congress.

Farms without electricity could not qualify as a Grade A dairy. Much milk was spoiled before it could reach the market. Photo courtesy of the Library of Congress.

Electricity first came to southwest Missouri to power the lead and zinc mines in the Joplin area. As early as 1890, electric motors began to replace the steam, mule, and manpower at the colorfully named mines such as the Rabbit's Foot, Yellow Dog, Grasshopper, Windy Bill, and Morning Star. The city of Springfield would have electricity as early as 1883 when the Springfield Electric Company was formed. Aurora got electricity in 1887, just two years after lead and zinc were discovered there. The lights were turned on for the first time in Mount Vernon on January 18, 1899.

To provide electricity, small companies were formed across the region. For many of these companies, electricity was just a sideline. Some of them existed more to produce ice than electricity. As early as 1887, Joplin residents could see the benefits of electricity when a small plant was built to power streetlights. A city ordinance passed that year gave the Joplin Gas and Coke Company the right to provide electricity to city residents.

Several smaller towns in the region—including Pierce City, Purdy, Neosho, Noel, Reeds Spring, Ash Grove, Willard, Galena, Walnut Grove, Stark City, Bois D'Arc, Strafford, and Fair Grove—all had electricity prior to the 1930s. In 1909, Joplin's "Electric Park," a ten-acre amusement park, was built. All of its attractions, including a tower, were draped with electric lights. Visitors could come to the park in an electrically powered streetcar.

On October 26, 1909, six small electric utilities merged into the Empire District Electric Company, named by its New York financiers in a nod to their own "Empire State." Power for the new company would come from the Lowell and Redings Mill hydroelectric plants, along with a generator at the Riverton Generating Station in

Students struggled to read and do homework by lantern light. They also had the chore of trimming lantern wicks and cleaning the soot from glass chimneys. Photo courtesy of the Library of Congress.

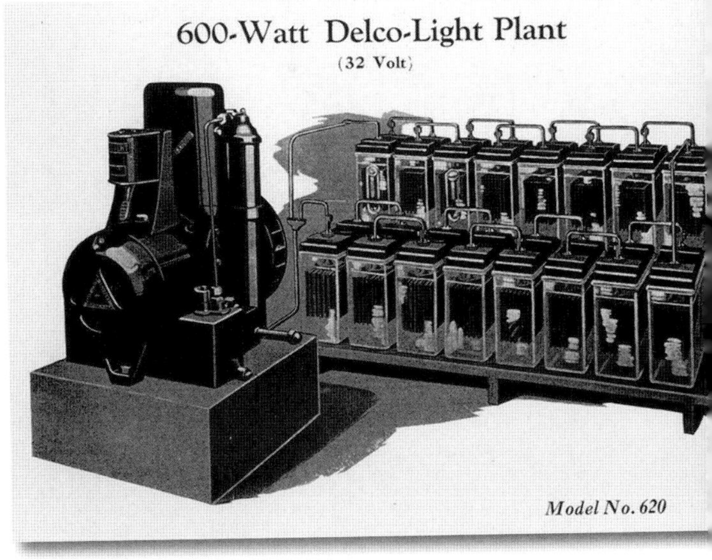

Tired of waiting for central station power, those families able to afford them bought Delco power plants. These kerosene-powered generators charged sixteen 2-volt batteries, enough to provide power for a radio or a few lights. Photo courtesy of Craig Toepfer.

Kansas. Empire also built Powersite Dam on the White River, which formed Lake Taneycomo and supplied 16 megawatts of electricity.

By 1922, Empire was providing electricity to 11,000 customers and thirty communities. But Empire and the smaller private power systems it did not acquire were only interested in serving in towns and cities where it was possible to make a profit. They could see no return on investment extending lines into rural areas where there were very few people per mile.

Rural people could see the lights on the horizon and they may have experienced electricity firsthand when they visited friends or relatives in the city. However, most were resigned to the fact that electricity was for city people. An article in *Popular Mechanics* magazine echoed that sentiment: "Thousands of these rural homes will never enjoy the blessings of electricity if they wait for the high lines to bring it because they are in areas so sparsely populated that power lines can not be made to pay for themselves." The article encouraged farmers to generate their own power using windchargers or generators powered by gasoline, kerosene, or acetylene.

Many farmers did turn to these devices, especially the 32-volt Delco power plants. Once a week, or more often if more power was needed, the units were started and run long enough to charge the sixteen 2-volt batteries. But the cantankerous machines broke down often, provided only enough power for a few lights or a radio, and were expensive to operate.

In the 1930s, only about 10 percent of the nation's farms had central station electric service. In Missouri, the situation was worse. Only one Missouri farm in fifteen had electricity. This condition stood out in stark contrast to the rest of the civilized world, where almost 95 percent of French farms, 90 percent of Japanese farms, and 100 percent of Dutch farms had electricity.

Cold in winter, hot in summer, and often plagued by wasps and snakes, the outhouse was a staple of rural living in the 1930s. Ads like this one encouraged rural people to build sanitary outhouses to protect public health. Photo courtesy of the Library of Congress.

Drought conditions and strong winds combined to create dust storms that darkened the skies and left many parts of the Midwest devoid of topsoil in the 1930s. The Dust Bowl days lasted from 1930 until 1936 and caused many farm families to leave the land. Photo courtesy of the Library of Congress.

In the words of Lorena Kemppainen, "We cared for and loved each other and did not realize that we were poor, even when we visited friends' homes where indoor plumbing was an accepted way of life. We had a simple goal—to get electricity so that we could improve our living conditions with lights and plumbing. Until REA became available, we simply accepted a situation we could do nothing about."

That task would begin in earnest as a result of one of the darkest periods in our nation's history. In 1929, the stock market crashed, and overnight millions were bankrupt and out of work. Out of the tragedy came a series of President Franklin D. Roosevelt's "New Deal" programs aimed at getting the nation back on its feet. One of these was the Rural Electrification Administration, or REA.

Roosevelt knew first-hand the plight of rural residents. Suffering from polio, the president vacationed in Warm Springs, Georgia, where the soothing waters eased the pain in his disease-stricken limbs. "There was only one discordant note in the first stay of mine at Warm Springs," Roosevelt said. "When the first-of-the-month bill came in for electric light for my little cottage, I found the charge was 18 cents a kilowatt-hour—about four times as much as I paid in Hyde Park, New York."

During the Great Depression, the Works Progress Administration provided relief work for unemployed men and training in public works such as building roads and parks. Photo courtesy of the Library of Congress.

That observation started Roosevelt's study of ways to get power into rural homes at rates farmers could afford. In his message to Congress in January 1935, he set the wheels in motion toward setting up a federal agency to help build rural power lines. On May 11, 1935, he signed Executive Order No. 7037, establishing the REA as an emergency agency "to initiate, administer and supervise a program of approved projects with respect to the generation of electric energy in rural areas." Relief funds were to be loaned to private power companies, municipals, and cooperatives.

A year later, REA became a permanent agency of the federal government, with $410 million to loan over ten years. Congressional leaders thought established utilities would leap at the chance to borrow low-interest money to build power lines. That was not the case. When the first applications came in, not one private power company was at the table. A special committee of these for-profit utilities studied the rural situation and came up with this interesting conclusion: "There are very few farms requiring electricity for major farm purposes that are not now served."

Instead, rural people, led by county extension agents, took matters into their own hands and formed electric cooperatives to finally bring electricity to rural people. Some said the task was too difficult for simple rural people who knew nothing about establishing, much less operating, an electric utility. Others thought electricity was a luxury they could not afford.

Undaunted, the famer took a hand in establishing a better future for his family. And out of this movement came Ozark Electric Cooperative. Southwest Missouri would have its lights!

Can We Afford Electricity?

W. L. Boucher explains how it can be afforded
From the December 9, 1937 *Lawrence County Chieftain*

William Boucher, one of the cooperative's original incorporators, was pressed into service as Ozark Electric's first project manager. He would lead the cooperative until January 1940, when personal matters demanded his attention.

Photo at left is courtesy of the Library of Congress.

Mr. and Mrs. Farm Folks, the question "Can we afford electricity?" is being discussed at this time in hundreds of farm homes, not only in your county, but in dozens of counties in this state.

Now let's see if you can afford not to have it. Let's not talk in terms of money, but in terms of farm produce. Do you know that seven eggs a day at twenty cents a dozen, will furnish electricity to do your washing, ironing, pump your water, operate your radio, light your house, barn, chicken house and garage? Is it worth it? All you have to do is turn your light on in the chicken house early in the morning and you can produce enough extra eggs to pay the bill. You can even arrange to have your lights come on without waking up to do it.

Shall we look further into what it will cost in farm commodities? One-third of a pound of cream a day at thirty cents per pound will furnish all of the above. Six pounds a day of whole milk, at one dollar and seventy-five cents per one hundred pounds, will furnish this service. One-ninth of a bushel of wheat a day at one dollar per bushel will give you these conveniences that you are paying for and not getting. Twelve crates of strawberries at a fair price will pay your electricity bill for an entire year.

For just a little more produce you can have your own refrigerator. Just think, a nice cold drink of water when you want it, ice tea or lemonade when you please, as well as a good dish of ice cream for your evening meal after a long hot day's work. You can grind your feed, saw your wood, operate the grindstone, fill the silo and a lot more odd jobs with a portable motor.

It is time to think about a little insurance for your vegetable garden and small fruits, such as raspberries, strawberries, etc. A small irrigation system from the well, spring or river will do the job. And why not figure a little further and say we will irrigate a small field of alfalfa or Atlas sorgo (sorghum used for silage) so that the old cow will have plenty to eat of what she needs and take that milk or cream check home or put it into the bank instead of paying it out for feed in these drought years.

Why not plan to produce just a little more farm produce and have a home in the country, just like the home Sister, Brother, or Mother and Dad have in town. You have this opportunity, why let it get away when it is going to relieve you of so much work, and make our city folk want to buy a home in the country.

Our land values have gone down, but we can bring them back with electricity available at the price we can afford to pay.

Each and every individual as well as the community either goes up or down, we don't stand in the same place long at a time. Rural electrification will enable us to progress faster than any one single development that we have been privileged to have in this generation.

Remember the more electricity used the cheaper it is.

Careful consideration and a little planning on our part will certainly prove that we cannot afford to pass up this opportunity to have electricity.

Switch Is Thrown And Power Surges To 125 Farm Ho

Here are the men who are responsible for the success or failure of the REA cooperative that urnished electric current to 125 families in the Ash Grove community. They are also the ones who the lines possible by getting the farmers to sign up for the current. The following, left to right, in ack row are members of the cooperative's board of directors: C. A. Sifferman, J. B. Williams, Harown, chairman, Justin Carlock, W. C. Cox, I. E. Murray and T. O. Holman. W. L. Boucher, manager of EA project in southwest Missouri, and County Agent C. C. Keller are seated in front of the board. the expressions on all the faces it seems that all are pleased with their work. They are seated in of the Empire District Electric company sub-station from which the cooperative gets its current.

When County Agent C. C. Keller shoved on the swit the "hot stick" as he is doing in the above picture, 125 miles of line received the first REA current turned county. The ceremony took place near Ash Grove at 1 terday afternoon. Those grouped around him, left to r Bucher, manager of the REA in this section; Harley Br of the Ash Grove district cooperative; H. L. Jones, mai E. R. Mead, resident engineer in charge of the line co O. J. Peugh, superintendent of construction

This newspaper clipping from September 1938 tells the story of when the Ash Grove line was energized for Ozark Electric. It also shows most of the original founders of the cooperative.

Chapter 2
Getting Started

On October 21, 1937, a diverse group of southwest Missouri men met in the Lawrence County Courthouse. The purpose was to form an electric cooperative to bring power to the countryside. Under the guidance of Fenton Stockard, an attorney who got most of Missouri's electric cooperatives underway, they would form a permanent organization that would become known as Ozark Electric Cooperative.

Under the laws of the state, Ozark Electric was first organized as a stock company. The original incorporators who signed the charter on October 20, 1937, were I. E. "Irby" Murray, Justin Carlock, Albert T. Renshaw, T. O. Holman, W. C. Cox, J. F. Hobbs, Forest C. Speight, J. B. Williams, Roy A. Sifferman, Arley Brown, C. A. Sifferman, and Don D. Lee.

Rules and regulations were adopted, and the board agreed to adopt bylaws as used by other REA organizations. At that meeting they unanimously agreed to hold the first annual meeting on the first Monday in February. Another important first step was to appoint someone to serve as project manager. W. L. Boucher was nominated and approved as Ozark Electric's first manager.

The next day, the first board of directors met. They were Arley Brown, Aurora; I. E. "Irby" Murray, Ash Grove; Justin Carlock, Everton; Forest Speight, Everton; Roy Sifferman, Republic; T. O. Holman, Stotts City; W. C. Cox, Aurora; J. B. Williams, La Russell; and C. A. Sifferman, Mount Vernon.

Officers elected that day were Arley Brown, president; I. E. Murray, vice president; and Justin Carlock, secretary/treasurer.

This organizational meeting was the culmination of several months of activity, beginning with meetings at the home of Irby Murray and later at a church in Ash Grove. On March 16, 1937, fifty people met in a countywide meeting held in the Lawrence County Courthouse in Mount Vernon. Those assembled heard K. B. Huff, an agricultural engineer from the University of Missouri, explain how rural people could form their own cooperative to provide electricity.

As the organizers moved from township to township holding informational meetings, they promised electricity could be had for as little as $3 a month. "It took a lot of selling," Murray would tell the *Monett Times* in a special edition printed in 1950. "We spent hour after hour convincing our neighbors that the REA would help us far beyond the cost."

Newspapers in the area helped with the effort. "It is entirely possible to have rural electrification in Lawrence County if the people want it," wrote the *Lawrence*

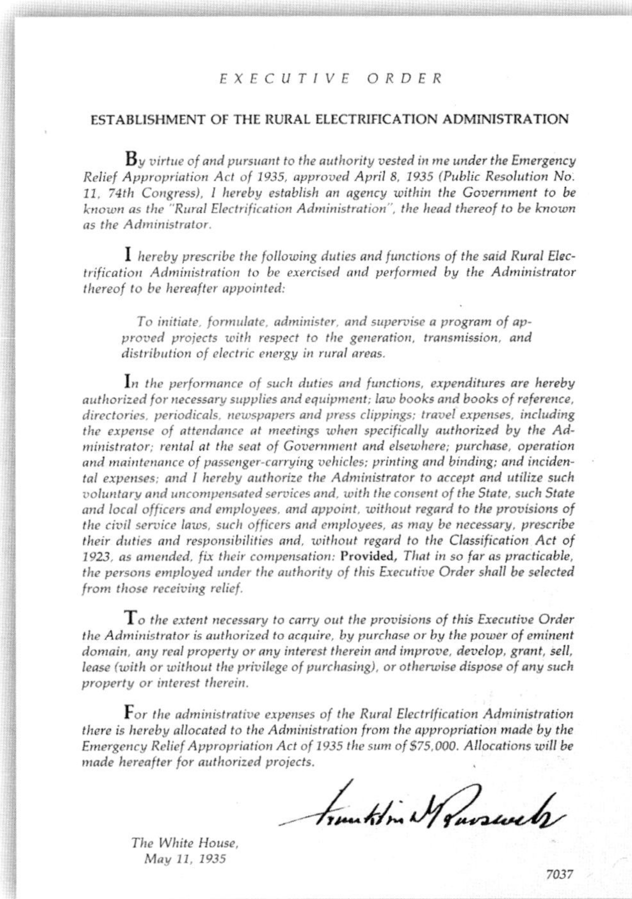

E X E C U T I V E   O R D E R

ESTABLISHMENT OF THE RURAL ELECTRIFICATION ADMINISTRATION

By virtue of and pursuant to the authority vested in me under the Emergency Relief Appropriation Act of 1935, approved April 8, 1935 (Public Resolution No. 11, 74th Congress), I hereby establish an agency within the Government to be known as the "Rural Electrification Administration", the head thereof to be known as the Administrator.

I hereby prescribe the following duties and functions of the said Rural Electrification Administration to be exercised and performed by the Administrator thereof to be hereafter appointed:

> *To initiate, formulate, administer, and supervise a program of approved projects with respect to the generation, transmission, and distribution of electric energy in rural areas.*

In the performance of such duties and functions, expenditures are hereby authorized for necessary supplies and equipment; law books and books of reference, directories, periodicals, newspapers and press clippings; travel expenses, including the expense of attendance at meetings when specifically authorized by the Administrator; rental at the seat of Government and elsewhere; purchase, operation and maintenance of passenger-carrying vehicles; printing and binding; and incidental expenses; and I hereby authorize the Administrator to accept and utilize such voluntary and uncompensated services and, with the consent of the State, such State and local officers and employees, and appoint, without regard to the provisions of the civil service laws, such officers and employees, as may be necessary, prescribe their duties and responsibilities and, without regard to the Classification Act of 1923, as amended, fix their compensation: **Provided,** *That in so far as practicable, the persons employed under the authority of this Executive Order shall be selected from those receiving relief.*

To the extent necessary to carry out the provisions of this Executive Order the Administrator is authorized to acquire, by purchase or by the power of eminent domain, any real property or any interest therein and improve, develop, grant, sell, lease (with or without the privilege of purchasing), or otherwise dispose of any such property or interest therein.

For the administrative expenses of the Rural Electrification Administration there is hereby allocated to the Administration from the appropriation made by the Emergency Relief Appropriation Act of 1935 the sum of $75,000. Allocations will be made hereafter for authorized projects.

The White House,
May 11, 1935

7037

President Roosevelt's Executive Order No. 7037 established the Rural Electrification Administration on May 11, 1935. Photo courtesy of the Library of Congress.

Roosevelt signed his historic executive order creating REA in this ceremony. The president understood the plight of rural people who did not have electricity due to the time he spent in Warm Springs, Georgia, seeking relief from his polio. Electricity here, though available, was four times what he paid in New York. Photo courtesy of the Library of Congress.

The Lawrence County Courthouse, shown here in 1937, was the scene for most of the early activity for Ozark Electric Cooperative, including where it was first incorporated. The cooperative's first office was located here until August 1938.

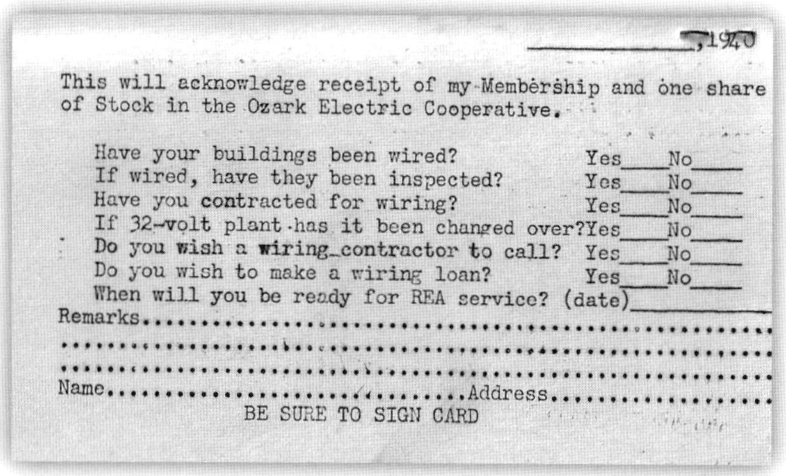

________, 1940

This will acknowledge receipt of my Membership and one share of Stock in the Ozark Electric Cooperative.

Have your buildings been wired? Yes____No____
If wired, have they been inspected? Yes____No____
Have you contracted for wiring? Yes____No____
If 32-volt plant has it been changed over? Yes____No____
Do you wish a wiring contractor to call? Yes____No____
Do you wish to make a wiring loan? Yes____No____
When will you be ready for REA service? (date)________

Remarks...

Name......................Address....................

BE SURE TO SIGN CARD

Those who applied for membership in Ozark Electric Cooperative were sent this postcard as a receipt of their membership fee. The questions helped REA determine when a line was ready for construction.

County Chieftain (now the *Lawrence County Record*) in its April 1, 1937 edition. "All that is necessary will be a little cooperation and working together of the people who want it. Three counties in the state have it now and six others are building lines at present."

The article explained that the Rural Electrification Administration would loan the cooperative money at 3 percent interest over twenty years. Required for the loan was an income between $10 to $18 per month over each section of proposed line to justify construction. "The minimum that the farmer will receive, between 30 and 60 kilowatt-hours, will furnish lights for the farm, power for the radio, washing machine, electric iron and other small electric appliances," the front-page news story related.

Getting the cooperative organized was probably the easy part. Ahead lay the daunting task of convincing enough members to sign on, and more importantly, part with the $5 membership fee.

"We met in rural schoolhouses and invited the farmers who were interested to come," said Raymond Phariss. "As I remember, the schools were pretty full. They wanted to learn about it."

He said the most interest came from dairy farmers and anyone who was in the chicken business. Dairymen saw the possibility of increasing herd size while at the same time cutting their workload in half with electric milkers and running water. Likewise, those raising poultry saw the benefits of plentiful water and electric fans to keep the chicks cool and electric lights under brooders to help them grow.

"We all knew it was extremely helpful," Raymond added. "But several thought it was a luxury they could do without. I know we went to one fellow who wouldn't sign up. He said,

'I don't know if I can pay that minimum.' I told him that we needed him to finish up this five miles. 'If you'll sign up, I'll pay your first dues. Then if you don't want it, you can unhook.' But I don't remember anyone who asked us to take them off."

With Boucher at the helm, the board wasted no time moving forward on what would be Missouri's most ambitious rural electrification project to date. Early plans called for building electric lines in Barry, Dade, Greene, Lawrence, Polk, and Stone Counties using a $150,000 loan from REA, a partial allotment from the $347,000 requested. These funds would build 325 miles of line to serve 906 members.

"We knew what we wanted and we stuck together until we got it," wrote Greene County Extension Agent C. C. Keller, who was instrumental in getting the cooperative off the ground. "We are going to have electricity."

Keller's letter dated November 24, 1937, was addressed to prospective members of the cooperative. It told of the loan from REA and cautioned that the coming months would be the hardest since progress would be slow. "Since we are definitely going to have electricity, it is now time to plan our wiring, decide on what appliances we want first and make provisions for hooking up to the line as soon as it comes. In this way we will get the full benefits of electricity at the beginning."

One of the first steps for the cooperative's board was to secure a supply for wholesale power. A contract was signed with Empire District Power Company, which had a surplus due to the closing of mines in the Joplin area. Empire's Riverton plant was one source of this power. Photo courtesy of Empire Electric.

Of course the project would need a source of wholesale power, and negotiations were underway to secure it from Empire Electric for an average of 1.25¢ per kilowatt-hour. In time, the project would take in everything from the southwest corner of the state to as far east as Webster County and as far north as Camden County. At the time, REA was urging cooperatives to merge in order to create efficiencies. In fact, REA demanded a merger at three central Missouri cooperatives that were considered too small to survive. That would not be necessary in southwest Missouri where directors envisioned one of the state's largest electric cooperatives.

No. 60975

STATE OF MISSOURI

Certificate of Incorporation

MDCCCXX

Whereas, An association organized under the name of OZARK ELECTRIC COOPERATIVE

has filed in the office of the Secretary of State Articles of Association or Agreement in writing, as provided by law, and has in all respects complied with the requirements of law governing the formation of Co-operative Associations for

CO-OPERATIVE AGRICULTURAL PURPOSES:

Now, Therefore, I, DWIGHT H. BROWN, Secretary of State of the State of Missouri, in virtue and by authority of law, do hereby certify that said association has, on the date hereof become a body corporate, duly organized under the name of OZARK ELECTRIC COOPERATIVE located at Mt Vernon and is entitled to all the rights and privileges granted to Co-operative Associations under the laws of this State for a term of perpetual years, and that the amount of the Capital Stock of said corporation is Thirty thousand---------------------- Dollars.

In Testimony Whereof, I hereunto set my hand and affix the Great Seal of the State of Missouri. Done at the City of Jefferson, this 21st. day of October A. D., Nineteen Hundred and Thirty seven.

Dwight H. Brown
SECRETARY OF STATE

CHIEF CLERK.

MIDLAND PRINTING CO., JEFFERSON CITY, MO. X6957

Ozark Electric's incorporation certificate was signed on October 21, 1937.

At the second board meeting, Boucher was given $250 for office supplies plus $75 a month to hire a stenographer. A typewriter would be purchased with payments extended over twelve months.

Justin Carlock was hired as easement director at a salary of $125 a month plus 3¢ per mile car allowance up to $35 a month. In order to increase the efficiency of the easement work, it was agreed to use the most outstanding individual in each township to assist in the work and to work in conjunction with the engineers.

Moving with lightning speed for such a project, the board had an engineering firm in place by its third meeting. They hired Sverdrup & Parcel from St. Louis, the same firm that had designed Sportsman's Park where the St. Louis Browns played, done work for electric cooperatives in Texas, and also built the fish hatchery at Lake of the Ozarks.

By November 29, 1937, rates were set: first 40 kilowatts, $3.25 a month minimum; next 40 kilowatts, 4.5¢; next 120 kilowatts, 3¢; all over 200, 1.5¢. An annual minimum charge of $24 per year for churches, schools, and community halls was set, along with $2 a month for non-farm customers.

Board members were tasked with doing the survey work. Each was given a section of the project and were paid $5 per day and had to furnish their own transportation.

In February 1938, Louvina Watterson, the new stenographer and bookkeeper, joined Boucher and Carlock. A month later three more women were hired—Edythe Hopper, Ione Standhardt, and Gladys McMahon.

By April 2, 1938, plans were approved by REA and bids were advertised for the construction of the first line. REA officials optimistically predicted members would see the current flowing by July. Boucher warned members that they needed to get homes wired in advance of the construction. They were given until March 26 to sign up for service and pay their membership fee to assure themselves of service. "If not signed up by that date, it is indefinite when you can be furnished service," the *Chieftain* warned.

Williams Brothers and Haas of Chicago and Tulsa was low bidder on the project. However a board resolution let REA know that the board preferred Neyer Construction of Billings, Missouri, whose bid was $107,203.12 versus $106,814.83 for Williams and Haas. The budget-conscious

REA gave the nod to the Chicago firm despite the board's local recommendation.

Conscious of the high unemployment in the area, Ozark Electric's board vowed to use local labor on the project wherever possible. A notice run in the March 31 *Chieftain* said that anyone wanting to work could sign up at the office in Mount Vernon, at Turk Brothers Store in Ash Grove, or at Forest Speight's store in Dade County. The notice brought out 250 men, of which 125 to 150 were employed, mostly to dig holes for the poles.

Now Ozark Electric needed a lineman to handle connects and do maintenance once the system was in place. Hubert Jones of Everton became the first maintenance man with a salary of $100 a month. The board also approved purchasing the cooperative's first truck in July 1938. They bought an International fitted with a McCabe-Powers utility bed at a cost of $275.

Work on Ozark Electric's lines began in earnest after REA approved its first loan in 1938. Williams Brothers and Haas of Chicago and Tulsa was low bidder on the project. The first line ran from Red Oak to Verona.

As work on the lines got started and members began to wire their homes in advance of the current, the cooperative secured a loan from REA that would allow members to install wiring and plumbing. This must have been welcome news at a time when cash for such improvements must have been hard to find. The loans would pay up to 80 percent of the cost and had to be paid back within five years.

The cooperative had been working out of the Lawrence County Courthouse, but as operations became more complex, REA officials encouraged the board to find office space elsewhere. Boucher found space in the Ray Richardson Building across from the Mount Vernon Post Office for $25 a month, and the cooperative moved in on August 12, 1938.

Meanwhile, extension agents from Lawrence, Greene, and Dade Counties sponsored meetings to teach members how to select and install electrical appliances and lighting. Excitement was building as the lines snaked out across the land.

By July 1938, seventy-five men were working in the Ash Grove area, clearing right of way, setting poles, and stringing wire. In August, an additional 450 miles of line was approved by REA as extensions off the main line, with the understanding they would be developed immediately. Additional counties added to the project included Christian, Cedar, Dade, Newton, and the eastern part of Jasper County.

With the workforce now at one hundred men, the Ash Grove to Dadeville branch was almost complete, despite considerable difficulty in digging holes in the rocky ground. In Lawrence County,

Early Ozark Electric linemen rode in this box built on the back of one of the trucks. Note the crosscut saw hanging on the box and the wooden barrels full of hardware.

The ladies of the office pool took care of new members, billing, and meeting minutes.

work was also progressing with poles being set as fast as possible. Plans were for work to continue into the winter on the new sections.

Finally the day everyone was waiting for arrived. "Rural Electrification becomes a Reality" was the headline on the front page of the *Chieftain* on September 29, 1938. The honor of throwing the switch that energized the first line on September 23, 1938, went to John W. Woodward, extension agent for Lawrence County. It was Woodward who called the first meeting and got the ball rolling for what became Ozark Electric Cooperative.

The first farm to receive electricity was owned by Don Lee, who lived near Verona. The first line energized ran from Verona to Red Oak on the west side of the project. It served just sixty-seven homes. But more ceremonies like this would take place, with the Ash Grove to Dadeville line being energized a week later and the rest of the project in October. A total of 137 miles of line was now energized, with enough money available to build an additional 1,200.

Many people still recall the "Day the Lights Came On." One of these is Don Wood, who retired after a long career with Ozark Electric and the Association of Missouri Electric Cooperatives. "In the 1940s the power came on, and I remember it well. My father and some of the neighbor men came to Mount Vernon to see if Ozark Electric Cooperative would set a meter at the Jackson School for a pie supper. When the men from Ozark Electric came out to set the meter they set ours too, because our home was to be served by the same transformer. I remember my father's excitement. As we were leaving for the pie supper, he said, 'Leave the pole light on so we can see to walk home!' "

Hazel Maze of Dadeville, who once held the record for the longest continuous

Until the loans came in, linemen drove their own trucks. This was Ozark Electric's early fleet.

Poles for the new lines came into Mount Vernon on trains and were unloaded behind the Carnation Milk Plant.

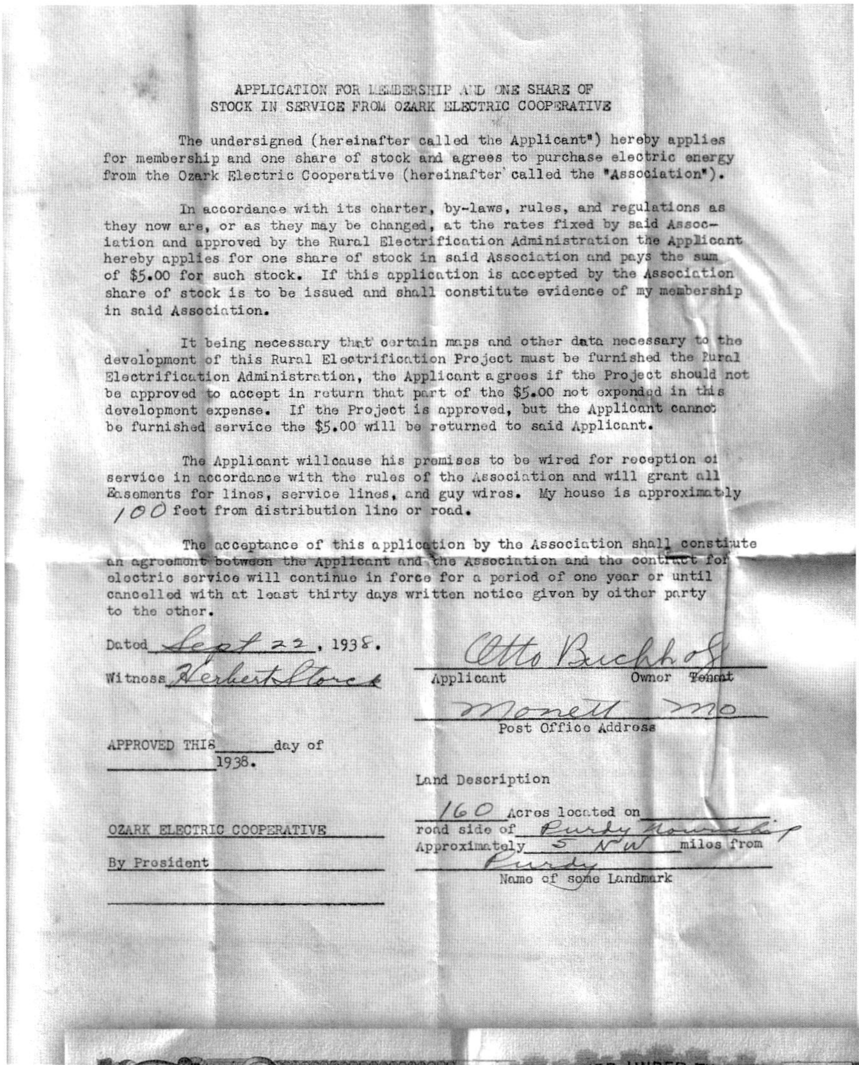

APPLICATION FOR MEMBERSHIP AND ONE SHARE OF
STOCK IN SERVICE FROM OZARK ELECTRIC COOPERATIVE

The undersigned (hereinafter called the Applicant") hereby applies for membership and one share of stock and agrees to purchase electric energy from the Ozark Electric Cooperative (hereinafter called the "Association").

In accordance with its charter, by-laws, rules, and regulations as they now are, or as they may be changed, at the rates fixed by said Association and approved by the Rural Electrification Administration the Applicant hereby applies for one share of stock in said Association and pays the sum of $5.00 for such stock. If this application is accepted by the Association share of stock is to be issued and shall constitute evidence of my membership in said Association.

It being necessary that certain maps and other data necessary to the development of this Rural Electrification Project must be furnished the Rural Electrification Administration, the Applicant agrees if the Project should not be approved to accept in return that part of the $5.00 not expended in this development expense. If the Project is approved, but the Applicant cannot be furnished service the $5.00 will be returned to said Applicant.

The Applicant willcause his premises to be wired for reception of service in accordance with the rules of the Association and will grant all Easements for lines, service lines, and guy wires. My house is approximately 100 feet from distribution line or road.

The acceptance of this application by the Association shall constitute an agreement between the Applicant and the Association and the contract for electric service will continue in force for a period of one year or until cancelled with at least thirty days written notice given by either party to the other.

Dated Sept 22, 1938. Otto Buchholz
Witness Herbert Storck Applicant Owner Tenant

Monett Mo
Post Office Address

APPROVED THIS ______ day of ______ 1938.

Land Description

160 Acres located on
road side of Purdy Township
Approximately 5 NW miles from
Purdy
Name of some Landmark

OZARK ELECTRIC COOPERATIVE

By President

Left: *When Otto Buchholz applied for membership with Ozark Electric on September 22, 1938, he filled out this application and paid a $5 membership fee, certifying his home was 100 feet from the proposed line. He owned a 160-acre farm near Purdy. Otto's great-grandson, Drew Meier, went through Ozark Electric's scholarship apprenticeship program and is now a journeyman lineman.*

Below: *This certificate for one share of stock in Ozark Electric Cooperative was issued to Otto Buchholz on November 20, 1939. It's signed by President Irby Murray and Secretary Louvina Watterson. No matter how wealthy they were or how much land they owned, each member received one share of stock and had one vote in the election of directors.*

INCORPORATED UNDER THE LAWS OF THE
State of Missouri
NUMBER 1058
SHARES 1
This Certifies that Otto Buchholz is the owner of
One Shares of the Capital Stock of
Ozark Electric Co-Operative of Mount Vernon, Missouri
Co-Operative reserves right to purchase this stock when offered for sale; this certificate not transferable when owner is indebted to Co-Operative (Secs. 4, 6, Art. 1, By Laws) and is transferable only on the Books of the Corporation in person or by Attorney on surrender of this Certificate properly endorsed.
In Witness Whereof the President and Secretary have hereunto subscribed their names and caused its corporate Seal to be hereto affixed at Mt. Vernon, Mo.
this Twentieth day of November A.D. 1939
Louvina Watterson SECRETARY
I. E. Murray PRESIDENT
SHARES $5.00 EACH

individual membership to Ozark Electric, also remembered that joyous occasion. "It was such a joy to know that you could punch a button and get light," she related, calling it the biggest single change in her lifetime. "I enjoyed it because it meant that I didn't have to fill the coal oil lamps every day."

A huge "REA Celebration" was planned for November 1, 1938, in Mount Vernon to mark the coming of rural electrification to southwest Missouri. In announcing the event, the *Chieftain* predicted it would bring the largest crowd ever to the city, since people from eight counties were about to get electricity for the first time. Eleven high school bands, along with drum corps, were invited to march in the parade. Demonstrations and contests were planned, including woodcutting, feed grinding, water pumping, egg beating, and cooking of harvest dinners.

Even Governor Lloyd Stark came to the celebration and spoke to the crowd. He was joined on stage by officials from REA, the state extension, and all of the counties included in the project. More than four thousand people took part in the celebration, which lasted all night.

That same day, the cooperative issued bills to those using the service. Ozark Electric's first wholesale power bill paid to Empire was just \$204.75. Boucher told the *Chieftain* that on November 15, the day the bills were due, not a single one of the 136 accounts was delinquent.

When the second meter reading took place, considerable growth in the use of electricity could be seen. Now there were 188 meters, and they used one-third more electricity than in October. A survey of appliances showed there were three milking machines, twenty electric motors, 105 radios, one electric range, fourteen refrigerators, seven vacuum cleaners, fifty washing machines,

Posters like this one generated interest in finally getting electricity on the farm. Photo courtesy of the Library of Congress.

The office of Ozark Electric moved from the courthouse to the Strand Theater (the Ray Richardson Building) in August 1938. Early annual meetings were held here also, and then members who attended were treated to a picture show as well.

three water pumps, 102 irons, two sewing machines, two hot plates, two battery chargers, and one air compressor.

The manager predicted Santa Claus would bring a lot more electric appliances to those who were connected. He also said plans were in place to extend service to 2,500 to 3,000 members over 1,000 miles of line in the coming year.

A new loan for $263,000 was approved before the end of 1938. This would build another 245 miles of line extending from Greene County to Cedar County to serve an additional seven hundred homes and businesses. Webster County was added to the project in a second extension that would cover 325 miles across the southern part of the system to serve nine hundred.

A third extension would take the cooperative's lines as far north as Camden County, making this the most ambitious rural electrification project in the state to date. "It is expected that once construction is begun, that it will continue through until the entire project has been built," the *Chieftain* reported. "This project promises to be Missouri's largest Rural Electrification Administration project. The total cost of the present project will be approximately $1 million, and will make electricity available to more than 5,000 rural families."

As the calendar rolled over to 1939, a new employee joined the cooperative. S. E. Roberts was hired to replace Hubert Jones as maintenance man. Later, Board President Arley Brown would resign, and Irby Murray moved up to president. In time, these two would lead Ozark Electric for many of its formative years.

Operations Manager Chester Swearingen poses with linemen who are installing regulators on the system, the first step in balancing the load once multiple lines were in place. He later became Ozark Electric's manager.

One of the chief organizers of Ozark Electric Cooperative was Irby Murray, a farmer from Ash Grove who would serve as the cooperative's second president.

This building became Ozark Electric's first real office building after it was completed in 1942. It was built on three acres of land donated by Mr. and Mrs. Hal Childress.

Who was on First?

As we look back on the history of Ozark Electric Cooperative after seventy-five years, it might be interesting to figure out who was the first member of the cooperative.

Ozark Electric had about three thousand members on line when the first memberships were assigned in the early 1940s. They were assigned in order after first alphabetizing the members by surname. Using this as a guide, the honor goes to Gene Abbot of Walnut Grove.

A history of the cooperative written in 1971 says the first man to sign up for electricity was Clyde Seward of rural Monett. The first farm to receive electricity was owned by Don Lee, who lived near Verona and was one of about sixty-seven members to receive power from Ozark Electric's first line, which was energized in September 1938.

The longest continuous connection to the cooperative is the Dadeville United Methodist Church. The church received power on October 1, 1938, and has continued as a member of Ozark Electric ever since.

The longest-serving director on the board today is Jimmie Crouch of Aurora. He was first elected to the board in 1972 and is one Ozark Electric member who can still remember the "Day the Lights Came On."

This specially designed International truck offered shelter for the men riding in the back and room for materials to be hauled to the job site.

Chapter 3
A Mighty Task

All of the new construction taking place at Ozark Electric had an interesting effect on rates. On April 1, 1939, REA approved lowering the rate paid to a minimum of $2.75 and $18 for churches and schools. This was the cooperative principle at work. In a cooperative, things came cheaper by the dozen. As more members signed on for service, the expenses could be spread over more people. Also, because the cooperative business model took profit out of the equation, rates could be set at just above the cost of doing business.

At the June 1939 board meeting, the first in a series of events occurred that would change the boundaries for Ozark Electric. Board member A. N. Walker asked for permission to resign from the board in order to join the board of newly formed New-Mac Electric Cooperative. Whether out of impatience with the progress of Ozark's lines or at the encouragement of REA, 384 people who were part of Ozark's Section C line in Newton County withdrew from Ozark Electric to join New-Mac.

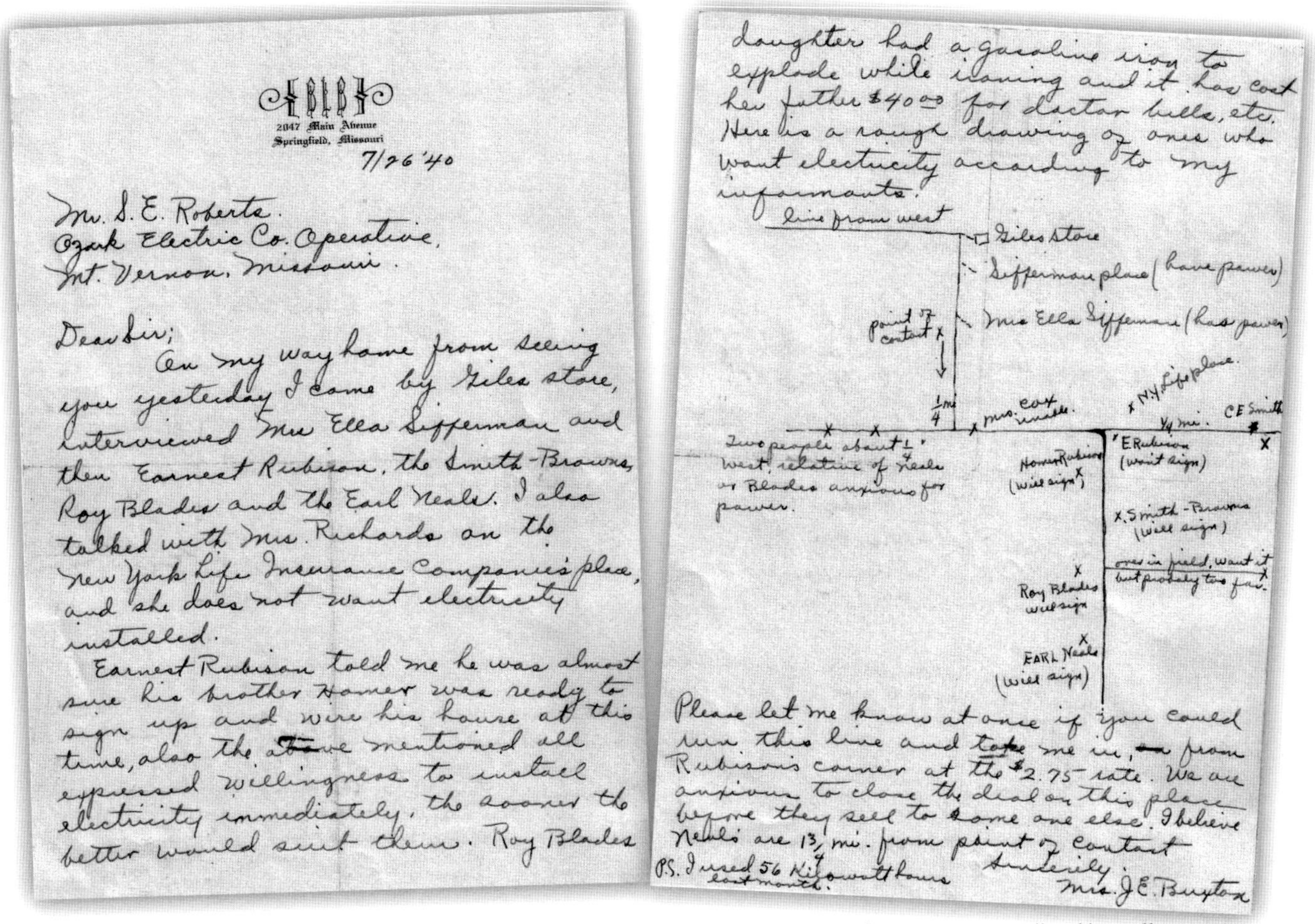

2047 Main Avenue
Springfield, Missouri
7/26'40

Mr. S. E. Roberts.
Ozark Electric Co. Operative,
Mt. Vernon, Missouri.

Dear Sir;

On my way home from seeing you yesterday I came by Giles store, interviewed Mrs Ella Sifferman and then Earnest Rubison, the Smith-Browns, Roy Blades and the Earl Neals. I also talked with Mrs. Richards on the New York Life Insurance Companie's place, and she does not want electricity installed.

Earnest Rubison told me he was almost sure his brother Homer was ready to sign up and wire his house at this time, also the above mentioned all expressed willingness to install electricity immediately, the sooner the better would suit them. Roy Blades daughter had a gasoline iron to explode while ironing and it has cost her father $40.00 for doctor bills, etc. Here is a rough drawing of ones who want electricity according to my informants.

Please let me know at once if you could run this line and take me in, from Rubison's corner at the $2.75 rate. We are anxious to close the deal on this place before they sell to some one else. I believe Neals are 1 3/4 mi. from point of contact

Sincerely,
Mrs. J E. Buxton

P.S. I used 56 Kilowatt hours last month.

Mrs. J. E. Buxton sent this letter to Ozark Electric Manager S. E. Roberts on July 26, 1940. It tells of her efforts to get more members signed up and also relates a story about the dangers of gas-powered irons, which had a nasty tendency to explode.

Ozark Electric's board agreed to Walker's resignation and also voted to return $1,136.64 to New-Mac. This was the membership fee for those members who transferred to the new electric co-op.

The New-Mac transfer was just the first of several. In October 1939, 493 members who were originally part of Section D were transferred to Southwest Electric, along with $1,545 in membership fees. It's possible that REA, or Manager Boucher, had decided the huge geographic area Ozark had carved out was unmanageable. Board minutes show frequent visits from members who wanted to know why line construction had stopped just short of their farms.

For example, Section B stopped 3.5 miles short because the REA loan money ran out before construction was completed. When Section D was approved, REA advised the board $155,000 was all that would be allowed on a low bid of $169,371. For this reason, twenty miles of line originally mapped had to be deleted from the section.

To help with management of the far-flung project, S. E. Roberts was promoted to assistant project superintendent. Two new maintenance men—Jimmie McClellan and John Pendelton—were hired. Also, the board approved requiring members to read their own meters at its July 31, 1939 meeting. This practice, common for electric cooperatives, would remain effective until the cooperative began installing automated meters in 2003.

But by January 1940, Ozark Electric's manager decided he could no longer stay on the job. He submitted his resignation as follows:

Gentlemen,

On account of personal business I find that I can no longer give all of my time to the business of Ozark Electric Cooperative. For this reason, I hereby offer my resignation to you and the Board of Directors to become effective on the acceptance day of this resignation. I have been closely connected with the development of this Project since the first meeting held at Mount Vernon, Missouri in March 1937 and I am proud to have had a part in the development of this Project and feel sure it will prove to be one of the greatest developments for Rural Southwest Missouri.

Wishing REA and this Project the very best of success, I am,

Yours very truly,

W.L. Boucher

Project Superintendent

Boucher would leave to manage Boucher's Hominy Plant in Monett. (He later shows up as manager of Crawford Electric based in Bourbon.) His resignation was reluctantly accepted, and S. E. Roberts, who had experience already in the electric utility industry, took over as temporary superintendent until the board could consider a replacement. That would take place on February 8, when Roberts was given the job permanently. A resolution states that he had a "thorough knowledge of the project and is qualified in every way to be superintendent." He was paid $150 per month.

His first duty was to squelch rumors that the home office was leaving Mount Vernon. The board passed a resolution at the 1940 annual meeting, the first held at the Strand Theater, that said: "We are thoroughly satisfied with the location of the home office of the REA at Mount Vernon, Missouri, and that there has never been any intention of even considering moving headquarters to any other place."

Shown here are four of the early Ozark Electric ladies—Sue Childress Compton, Evelyn Reece Marr, Nettie Stauffer, and Juanita Fulton.

With construction proceeding at a rapid pace and members coming onto the lines by the hundreds, the board asked REA for another wiring loan of $5,000, its third such loan. Also at the advice of Miss Agnes Watson, who worked for REA's utilization division, the board approved buying roasters and other appliances to be used at kitchen demonstrations. She approved of other plans in place to promote refrigerators, feed grinders, and small appliances during the holiday season.

Clearly, Ozark's board and management understood they could not simply build lines, connect members, and hope for the cooperative to prosper. Rural people unaccustomed to the benefits of electricity needed to be shown how it could work for them.

Ozark Electric helped two more cooperatives get started in 1940. In August, members on the north end of the system were transferred to Sac-Osage Electric Cooperative at El Dorado Springs. The next month, members on the southern end of the system were transferred to White River Valley Electric Cooperative based in Branson.

While these infusions of new members were small, they made a big difference in getting the new cooperatives off the ground. White River Valley, for example, had earlier failed to get enough members for the project to get started. A second effort succeeded, but only after REA was convinced the meager $400 average income of the hill folk was supplemented by their self-sufficient ways. When it came to satisfying a far-off banker that wanted to ensure its loans got repaid, even a hundred former Ozark members made a difference.

S. E. Roberts served as Ozark Electric's manager twice. In between, he helped organize Sho-Me Power, a transmission cooperative based in Marshfield, Missouri.

With Ozark on solid ground financially, and with 2,562 members receiving service, the board decided it was time for a real office. REA officials paid a visit in the summer of 1940 and liked what they saw financially. They encouraged the co-op to install a sign, preferably an electric one. In September 1940, representatives of the Mount Vernon City Council, Rotary Club, and Commercial Club met with the board and offered to donate a lot and $1,000 to help with a new office.

Instead, the board accepted the donation of three acres of land north of town from Mr. and Mrs. Hal Childress, along with a $1,000 donation from the Mount Vernon Ad Club. With approval from REA, work on the office began with architect Bruce Granger from Springfield doing the design. With several additions, this site remains the location of Ozark Electric's main office today.

Sue Childress Compton, whose father donated the headquarters land, later became a work order clerk for Ozark Electric. She recalls the land being nothing but "a blackberry patch" before the cooperative took over. "My dad and Mr. S.E. Roberts were friends for a long time," she says. "He was looking for a location for the office. My dad just gave it to him for a dollar. It was just a little old place they were renting. They had to have a bigger place."

Once settled into the new headquarters, the work of expanding the cooperative toward its goal of area coverage began in earnest. A request for a loan of $35,000 from REA was made in order to finance installation of indoor plumbing for three hundred members. Instead, REA offered $40,000 so more members could give up their outhouses once and for all. In September 1941, the board hired

Frank Horton & Company to install oil circuit reclosers in order to balance the load of the rapidly growing system.

With plans for Section F approved, the cooperative had enough money in the bank to make a $15,000 advance payment on its REA loans under what was known as "cushion of credit." These pre-payments were a hedge against a bad year when the cooperative might not be able to make its payments.

But just as Ozark Electric began moving full speed ahead on plans for area coverage, an event in a far off land would bring progress to a screeching halt. On December 7, 1941, Japanese planes bombed Pearl Harbor. In short order, the United States was at war with Japan, Germany, and Italy.

Rural electrification took a back seat to the war effort. Materials were scarce due to the war, with the government deciding that copper wire once intended to power the countryside was instead needed for tanks, planes, and battleships. Poles, wire, and other materials went first to the war effort. Only utilities that could prove war necessity got materials.

Construction came to a virtual standstill at most electric cooperatives and Ozark was no exception. The fledgling cooperative quickly found itself in a strange and ironic position. Farmers nationwide were being asked to do everything possible to become more productive so the nation could feed its hungry army. The one thing farmers needed to become more productive—electricity—would be denied them because copper wire was deemed more important for war materials than for rural electric power lines.

Quotas were set up based on farm "units." To get materials, a farm had to have so many units per mile. For example, one milk cow equaled one unit. Ten beef cattle equaled one unit. Hogs and chickens had similar values, but row crops didn't count.

Original Ozark Electric headquarters building after red and white awnings and a neon sign were added.

Ozark's board, elected by the members, had the daunting task of setting policies for the cooperative's operations, no easy feat for a group of farmers. Shown here are, clockwise from left: Manager A. Claypool Daugherty; President Irby Murray, Ash Grove; Albert Mason, Mount Vernon; George Marmon, Nixa; Van Anderson, Morrisville; C. C. Sherman, Rogersville; Vice President Hobart Little, Clever; Charles L. Moore, Mount Vernon; Secretary/Treasurer M. Henry Kemp, Golden City; and Edward Staley, Willard.

If a farmer didn't have enough units to qualify for electricity, he would borrow some livestock from his neighbors to meet whatever the requirements were at the time. They would drive herds all over to get the units they needed.

Any project that was 40 percent complete as of December 5, 1941, could be finished. All new projects were put on hold unless they were considered essential to the war effort or considered necessary for public health. Within three months all work on REA lines was halted regardless of its status. However investor-owned utilities were allowed to continue work. Frustration out of this obvious double standard led to the formation of the National Rural Electric Cooperative Association (NRECA) in March 1942, which Ozark Electric would join in September 1942.

That same month the cooperative joined the Safety and Training Program of the Missouri Rural Electrification Association, now known as the Association of Missouri Electric Cooperatives. This early program of the statewide association sent safety instructors to the distribution cooperatives to teach line crews safe working techniques. This was the first formal training many of the linemen received.

Many of those who had electricity were featured in the Honor Roll of those using the most kilowatt-hours. This listing was published in the cooperative newsletter. By August 1942, only those using 100+ kilowatt-hours were featured. Boyd Fugitt led the pack, using 1,180 kilowatt-hours.

Besides suffering from shortages of materials, the cooperative also lost employees to the war effort. One of these was lineman Joe Marshall, despite a trip by the manager to Jefferson City to

Once construction got underway, lines began snaking across the cooperative's extensive service area, one of the largest in the state. This crew changes out a damaged crossarm without interrupting service. Shown on the ground are Charlie Fender, John Coats, and James (Shorty) Barnes. On pole are Gene Matthews (bottom) and Doyle Warden (top).

appeal for a deferment. When he entered the service in 1943, the board wrote a letter of appreciation for his stellar work at the cooperative. His replacement was Chester Swearingen, another person who would play a big role in Ozark Electric's future.

Ozark Electric lost another key employee in January 1944. Reporting on the latest meeting of Sho-Me Power, Manager S. E. Roberts said he had been asked to take over as its manager. Roberts was given a leave of absence for one year and board member George Robbins became acting manager for one month. But a month later, Roberts asked for his unconditional release, which was approved.

A search for a new manager was launched and the board's first choice was Boone Electric Manager R. A. Dinwiddie. Perhaps Boone Electric offered more to keep him than the $300 a month Ozark was offering. Dinwiddie would stay in Columbia, and Ozark's board instead hired Claypool Daugherty as its new leader.

Before coming to Ozark Electric, the soft-spoken manager taught school for eight years, worked a year for the Soil Conservation Service, and was then hired as manager of Barton County Electric Cooperative in Lamar, where he worked for seven years.

At the time, Ozark Electric had a staff of eleven employees to care for 1,002 miles of line and 2,509 members. The typical member used 57.8 kilowatt-hours and paid a monthly bill of $3.84. With a density of 2.5 members per mile of line, that amounted to only $9.61 in revenue per mile.

One of those eleven employees was Mary Ann Moore Creed, who was hired to be the manager's secretary. "He hadn't had a secretary for six months, and when I went in his office there was a stack of correspondence a foot high he hadn't answered," she recalls. "Everyone was wanting service. Some of them got pretty nasty. Somebody down the road would get it and it wouldn't run to their house. Of course we had trouble getting material and supplies that we needed. Copper was almost impossible to get."

During World War II, the only progress the cooperative made was short line extensions as allowed by the War Production Board. Ozark Electric bought war bonds and used its mail list to notify members of a scrap iron drive.

This was Ozark Electric's staff in 1947.

Membership increased slowly to 3,236 by April 1943. Most of this growth came from the acquisition of several small towns from Sho-Me Power. The cooperative was now serving members in ten counties: Barry, Cedar, Christian, Dade, Greene, Jasper, Lawrence, Polk, Stone, and Webster. It had borrowed $919,000 from REA.

The annual meeting for 1944 was cancelled, and all directors were authorized to take care of right-of-way problems in their area. Ray Foree, a contractor based in Lamar who was hired to complete work on one section of line, reported the problems he was having to the board in August 1944. He said the material situation was critical, and applications for service were coming in faster than they could possibly be taken care of. At that meeting, the manager reported 175 work orders scattered out evenly across the system, covering 160 miles of line. In one month, Ozark employees had made seventy new connects, eleven disconnects, and changed out thirteen meters damaged by lightning. There were thirty-nine outages and ten transformers had been lost.

Manager Claypool Daugherty got in the spirit for the 1947 Lawrence County centennial with this Ozark Electric parade entry. Most of the men grew beards for the event.

Ozark Electric used a variety of locations for its annual meeting. This circus tent was provided by the statewide association, then called the Missouri State Rural Electrification Association.

Another of Ozark Electric's vehicles was this panel truck. Shown, from left: Ralph Williams, Orville Ross, Chester Swearingen, Wayne Walker, and Claude Dennison. This photo was taken in 1947.

In the 1940s REA encouraged electric co-ops to build community food lockers where meat could be stored frozen for later use. Ozark Electric built this food cooperative , Mt. Vernon Electric Refrigeration Cooperative, Inc., and helped it stay afloat when it could not make loan payments. The locker plant was the third highest user of electricity for many years.

A total of 185 miles of line was underway but unable to be completed. Ozark Electric had no transformers left even if it did get the lines completed.

In response, the board agreed to write letters to Congress saying Missouri needed more REA money allotted this year, not next year. But the progressive board didn't stay idle during the war years. Instead, they approved a project that was of great benefit to the cooperative's members despite being a money loser for many years.

The project was a community refrigeration cooperative, designed to reduce spoilage of meat for residents who had not yet been connected to the power lines or could not afford to buy a freezer. A board resolution read: "In view of the fact that the area around Mount Vernon, State of Missouri, is a dairying community and such type of farming requires much labor, it is believed that a freezer locker plant with slaughtering and processing equipment would materially help to increase the amount of foodstuff for consumption in this community and also help to alleviate the labor shortage; and it is further resolved that the Board of Directors of Ozark Electric Cooperative will guarantee the repayment of the loan from the United States of America to the Refrigeration Cooperative."

The resolution applied to food lockers in Mount Vernon and Marionville. Records show the Marionville plant was built during the war, but Ozark Electric was not actively involved in it. The Mount Vernon plant was a different story. Besides guaranteeing the loan, Ozark Electric donated an acre of land next to the office for the plant and its assistant manager, Marvin Carver, headed the board. The locker opened in September 1945 and cost $45,000 to build.

To secure a locker for storing meat at the cooperative plant, a person had to pay a $10

Outside employees gathered for this group shot in the late 1940s.

membership fee plus monthly rental. Besides storing food for members, the plant offered custom slaughtering, processing, and packing.

The food locker never proved profitable, and the board would have to make good on its promise to make loan payments when the plant was unable to generate enough money to pay its debts. Eventually, Ozark Electric bought the note from REA and operated it for a time, seeing a small return on its investment. Eventually it was sold.

Its real value, however, was in the large amount of power it consumed. At one time, it was the third largest consumer of electricity on the system.

As 1944 ended, contractor Ray Foree reported the material situation had eased and enough poles had arrived to complete the remaining 134 miles he was working on. He was given another contract to build an additional eighty-five miles of line to serve 227 members.

A 1945 annual meeting was a must as the cooperative was nearing the debt ceiling members had approved. The new manager would ask for the limit to be raised from $1.5 million to $5 million in anticipation of massive post-war construction.

He headed to St. Louis, headquarters for REA during the war, to request funds to build a garage, purchase a new truck, and make repairs and additions to the office. As the war drew to a close in Europe, an audit showed Ozark Electric had $48,781 in loan funds available, $16,100 in general funds, $15,340 in government bonds, and $35,697 in its cushion of credit account. Daugherty wanted funds to build 122 miles of line not in the Foree contract, along with a garage and warehouse. He also requested $10,000 for transformers, $4,000 for trucks, $4,000 for office equipment, and $500 for tools and other equipment.

On top of this came a loan of $150,000 to build lines to serve 463 new members and a complete system study to ensure area coverage was achieved. Poised for dramatic post-war growth, Ozark Electric released many of its members in Barry County to the newly formed Barry

Claypool Daugherty uses a Dictaphone to record letters for later typing. He became Ozark's manager in 1944 after serving as manager of Barton County Electric in Lamar.

Cashier Nettie Stauffer is shown in the foreground in this photo of the front office at Ozark Electric. With her are (counterclockwise) Evelyn Reece Marr, Juanita Fulton, and Mary Ann Moore Creed.

Electric Cooperative. Ozark also lost board member George Robbins, who became branch office manager for New-Mac Electric Cooperative's Cassville office and later manager of Barry Electric.

The board passed a resolution calling for cooperation with any new cooperative formed by REA. This would quickly be tested as Webster Electric Cooperative was formed. Once again, Ozark transferred members and lines to the new cooperative, helping yet another electric co-op get off the ground. By 1946, all lines in Polk County were sold to Southwest Electric and lines in parts of Dade and Cedar Counties went to Sac-Osage Electric. In both cases the price for the lines was the cost of their construction plus 10 percent.

The system study showed much work remained to be done, and frustration ran high from the cooperative's board, employees, and members. Many who signed on before the war started had been waiting for five long years to get electricity. The study showed 1,800 miles of line would be needed and approximately three thousand area families were still without electricity. Instead of the $5 million loan ceiling originally believed to be sufficient, the amount approved at the 1945 annual meeting was doubled to $10 million.

By 1946, the number of members had grown to 4,457. In this year, a special meeting of the members was called to convert the cooperative to a non-profit membership cooperative. Each previous stockholder in Ozark Electric now became an official member.

Now the board would challenge President Harry Truman, whose order to the Department of Agriculture to slow down the operations of the Rural Electrification Administration, had raised the hackles of National Rural Electric Cooperative Association President Clyde Ellis.

This aerial view of Ozark Electric's headquarters shows the office and warehouse, along with the food locker to the left, in 1947. The headquarters is still located here at the intersection of Highways 39 and 174 north of Mount Vernon.

Not long after the first lines were built, the need to keep right of way clear became apparent. This tractor was converted for spraying brush in the right of way.

Linemen loved to pose with their trucks. In this photo, they were joined by President Irby Murray (kneeling left) and Manager Claypool Daughtery (kneeling right).

Truman believed REA expenditures should be held to a minimum consistent with the curtailed activities of other government agencies following World War II. In response, the Ozark board passed this resolution:

> BE IT THEREFORE RESOLVED, that the Cooperative endorses the general principle of curbing inflation as indicated in President Truman's order requiring the cessation of public works but respectively suggests that attention of the President as well as all concerned, be called to the fact that the Cooperatives of the Nation should not and cannot be classified as public workers and that the REA program is and has been, a great factor in producing of food products; that the labor situation is critical; that the agricultural districts must naturally turn to machines operated by electricity rather than human labor and the curtailment of the rural electrification program and the extension thereof will curtail the production of food and food products. Many farmers are threatening to disburse their herds of milk cattle unless they can be furnished electricity so as to operate by proper machinery. This only illustrates the importance of continuing the extension of REA and we respectively suggest that the order be rescinded in as far as it applies to the extension of rural electrification.

The combined resolve of Ellis and the electric cooperatives across the country won the day. By September 1946, word came down from REA that line work could continue but funds for new trucks and equipment would be delayed. At the October 1946 board meeting, Daugherty related, "The co-op needs new trucks as the old ones are wore out and requiring considerable expense to keep them running. However, trucks are not available at this time."

So desperate was the cooperative for new vehicles that extreme measures were taken. One prospective member had a heart condition. He told the cooperative he was certain he could secure a new truck for the cooperative if they would return the favor by completing his service. Since REA had ordered that disabled people were to be given priority, a motion was passed telling the manager to proceed in any way he saw fit.

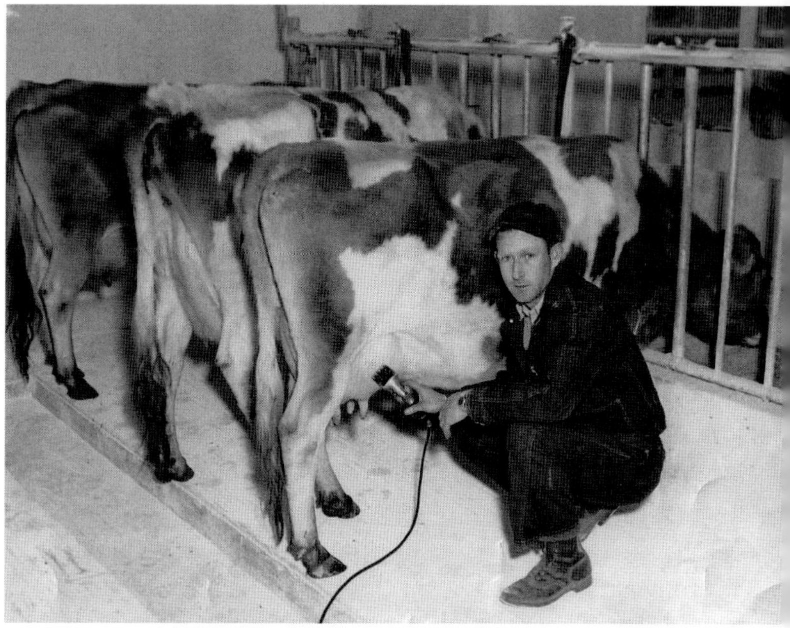

Electricity created new opportunities for Ozark Electric's many dairy farmers, including Joe Williams, an Ozark Electric member who lived north of Halltown.

Another motion in 1947 called for the establishment of a power-use department along with hiring a public relations person. In addition, the board approved sending the *Rural Electric Missourian* statewide newspaper (now called *Rural Missouri*) to all members once it began operation. A home economist would also be hired and would use kitchen facilities at the American Legion to host cooking demonstrations.

In May 1948, the cooperative purchased its first two-way radio. This would speed up outage restoration considerably now that crews in the field could be contacted directly.

As the decade closed out, members who once had to be begged to sign up for service were now begging the cooperative to get them power. More than one thousand were signed up for lines that had yet to be funded by REA.

Mr. and Mrs. George McKinley of Aurora appeared before the board to request service in time for their fiftieth wedding anniversary celebration. Mrs. H. A. Thomas read a letter from her doctor saying she needed electricity to help with her asthma. Another area farmer, Arthur Harter of Billings, could no longer work his dairy due to myocardial damage. The board agreed to build him a line as soon as possible, and in all other cases was sympathetic to the special requests for service.

Ozark Electric member Stephen Wright of Hurley recalls a story that confirms this "cooperative compassion." During World War II, his dad, Harold Wright, and uncles, Frank and Elton Wright, were killed. In 1948, the bodies of all three were returned to southwest Missouri for burial. Stephen's grandfather, Henry Wright, received a call from the cooperative asking what could be done to help in the wake of this tragedy. "He said he would really like to have electricity before the

For a time, the annual meeting was held in the warehouse located behind the office. Members who attended were treated to a meal.

bodies were returned," Stephen said. "They were just starting to string line down Highway A east of Hurley." The cooperative quickly built a line to the family as an act of compassion.

Stephen added, "Most of the people in this area, as soon as they got electricity they bought a Frigidaire refrigerator, always that brand. They had a light on a pull chain in each room. That was it. They went as cheap as they could and worried that the light bill would cost them more than kerosene. It was 10¢ a gallon. That was cheap light."

In 1948, REA made one of the largest loans ever to Ozark Electric, a whopping $1,150,000. The money would pay for 440 miles of line to serve 1,466 members, to make improvements to existing lines, and to pay for shortfalls in other sections hit by rising construction costs.

By April 1949, Ozark Electric had worked its way through the entire alphabet in naming line sections. The system had grown from 137 miles of line serving 149 members in 1938 to 1,830 miles of line serving 4,868 members in 1948. In 1949 there were plans for building 468 additional miles of line to serve an additional 1,405 members. In 1950, plans called for 435 miles of line to serve another 1,305 members.

Despite the all-out construction effort, which was now costing twice the $775 per mile paid in 1938, members were paying less for electricity than they did when the first lines were energized.

In celebration of all the hard work, the board passed a resolution directing the board and key employees accept George Marmon's invitation of spending the weekend at his cabin near Shell Knob for the purpose of fishing. "And be it further resolved that Mr. Marmon is hereby respectfully urged to begin preparations for such an outing in May at the time the fish are biting," the minutes relate.

Robbins Planned to Farm Again But Couldn't Stop Building Lines to Neighbors

Reprinted from the May 1954 *Rural Electric Missourian*

Back in March 1938, George N. Robbins, a Barry County farmer from Monett, was attending the funeral of a neighbor. It was a chilly day with snow blanketing the area. An Ozark Electric truck drove up.

"Are you George Robbins?" the driver asked.

"That's right."

"Well, you have just been elected to the board of directors of the co-op and you are wanted at the business meeting."

It was a surprise to Robbins, who hadn't even been a candidate for the post. At the business meeting he was elected secretary-treasurer of the Mount Vernon co-op—an office he held for nearly seven years.

In 1930, young Robbins decided he would like to have electric service on his farm. He lived one-half mile from the distribution line of Empire Electric. Between his house and the power line lived two other farmers. They wanted electricity too.

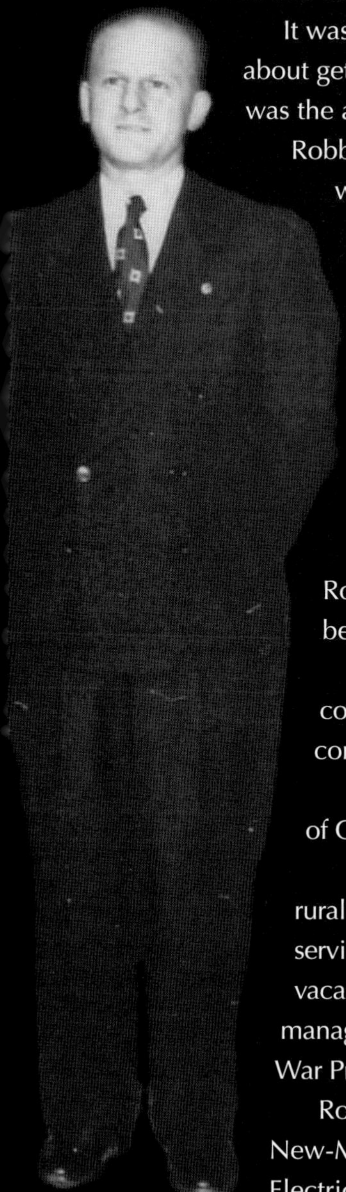

It was agreed that Robbins would call at the branch office of the company and enquire about getting service. "We will build the half-mile line for $1,000," or words to that affect was the answer.

Robbins and his neighbors didn't have that kind of money. "Go and get a home plant," was the advice given, as Robbins tells it.

Time marched on. The REA program was authorized in Washington. Empire Electric began building rural lines feverishly, Robbins related. Contacting the firm's development man, Robbins again sought service.

This time the story was different. The one grand—just forget about that. We'll be out right away, Robbins said he was told.

Again time went by and nothing happened. George went back to the branch office to check up on his application. He said that the Empire representative phoned the Joplin office of the company and discussed the matter.

"He hung up the phone and told me he couldn't do anything for me," Robbins recounted. "I was told to go up to Mount Vernon where an REA co-op was being organized and get service there."

"He said the co-op lines would be connected to Empire's system and that the company was going to let the farmers build the lines and make a failure so the company could take them over at 50 cents on the dollar," Robbins recalled.

"That's just what I did," George said. "I got in touch with W.L. Boucher, manager of Ozark Electric Cooperative. He came down and held a meeting in our community."

Robbins volunteered as a community leader to travel from farm to farm acquainting rural people with the new REA program and helping them prepare applications for service. Late in 1942, the manager resigned, and Robbins was again drafted to fill the vacancy until another manager could be employed. After serving three months as acting manager, he was assigned the job of helping labor-short farmers qualify for service under War Production Board animal unit regulations.

Robbins tried to return to farming, but was again brought back to REA work when New-Mac Electric tapped him to head its Cassville office. He became manager of Barry Electric when that system was organized in 1945.

Switching stations like this one located at Gore, Oklahoma, allow KAMO to isolate sections of its transmission grid for repairs without cutting power to the entire network. Photo courtesy of KAMO Power.

Chapter 4
KAMO to the Rescue

Another historic meeting took place in the spring of 1941 as S. E. Roberts and Irby Murray took part in the first meeting held to organize KAMO Power. When KAMO was formed, it marked the beginning of the end for rural power shortages.

Early in its history, Ozark Electric purchased its wholesale power from Empire District Electric Company, a competitor that did not necessarily have an interest in seeing the upstart electric cooperative succeed. Ozark was fortunate in getting started at a time when many of the mines Empire supplied power to were shutting down. The investor-owned utility had plenty of power to sell, at least when members were few and they considered a single light bulb hanging from a wire in the ceiling sufficient.

KAMO honored three of its founders and original officers at the last board meeting they attended in October 1953. Shown here are, from left: President John Hobbs, Vice President Irby Murray, and Secretary Dan Chandler.

However, as was the case with all of the other cooperatives in the state, Ozark's power supplier had its own customers to serve, and these folks came first whenever a shortage of power occurred. With cooperative members across the state using more and more power, it was just a matter of time before a cooperative source for wholesale power was discussed. At first Sho-Me Power was considered as a statewide cooperative providing wholesale power for all the state's electric cooperatives. But later a regional plan for six generation and transmission (G&T) cooperatives was seen as a better fit for Missouri.

One of the first of these to get underway was KAMO Power, originally named for member systems in Kansas, Arkansas, Missouri, and Oklahoma. Ozark Electric bought a $100 membership in KAMO on May 2, 1941.

Ozark also worked with Sho-Me Power to provide service to the cities of Fordland, Rogersville, Diggins, and Seymour. These towns were served with a minimum rate of $1.25 for the first 15 kilowatt-hours. These towns were added to Ozark as a part of Sho-Me's purchase of the electric service in various towns across the Ozarks from the Missouri Electric Power Company.

This purchase required approval from the Missouri Public Service Commission. Ozark's board passed a resolution in support of the purchase with the understanding that Sho-Me would then sell

This tanker truck kept KAMO's heavy equipment supplied with fuel. Photo courtesy of KAMO Power.

Irby Murray is shown here in front of the Ash Grove substation where power is transferred from KAMO's transmission lines to Ozark's distribution lines. Murray played a big role in the organization of KAMO and served as its vice president in addition to being Ozark's board president.

some of the lines it was purchasing to Ozark. On September 26, 1942, Ozark also became a member of Sho-Me Power at the suggestion of REA.

Ozark would provide Sho-Me with a manager when S. E. Roberts left in January 1944. He would serve the power supplier with distinction for ten years. But ultimately, Ozark would cast its lot with KAMO as its power supplier. A page-one headline in the first *Rural Electric Missourian* summed up KAMO's reason for being: "Lower Rates Aim of KAMO:" "Steps for bringing low-cost electricity out to the farms and rural industries of a four-state area which includes southwest Missouri were the main topics of discussion at the annual meeting of KAMO Electric Cooperative at Vinita, Okla., Nov. 20 (1947)."

KAMO got off to an early start in its efforts to bring power to Kansas, Arkansas, Missouri, and Oklahoma. Unlike the other G&Ts, KAMO did not intend to build its own power plant. The plan was to sell power produced by Oklahoma's Grand River Dam Authority (GRDA), created by the Oklahoma legislature in 1935.

GRDA was charged with harnessing Oklahoma's Grand River by building dams to produce hydroelectric power and for flood control. Within three years, the dream of a mighty dam across the Grand River was becoming a reality.

Thousands of men and women flocked to the construction site between the sudden boomtowns of Disney and Langley, looking for work during some of the darkest days of the Great Depression. Together, they built one of the world's longest multiple arch dams, stretching for more than a mile across the river.

A large crowd was on hand to help celebrate the official hand over of the Pensacola Dam to the Grand River Dam Authority following World War II. During the war, power from the dam was used for war industries but could now be used to provide power to KAMO member electric cooperatives. Photo courtesy of Grand River Dam Authority.

By the end of 1940, Pensacola Dam was in full operation, with the KAMO electric cooperatives hungry for its power. But World War II put the project on hold, with the War Department taking the electricity from the Grand River Dam Authority to operate an ordnance plant.

During the summer of 1941, KAMO signed contracts for the construction of three sections of 44,000-volt transmission line, a total of 348 miles. By March 1942, 188 miles of poles, complete save for the copper conductor, had been erected. All of the hardware, poles, insulators, and crossarms necessary for completing the project were on hand. But the project was not completed due to diversion of wire to the war effort.

During World War II, the power cooperative maintained only a part-time staff based in Hollister, Missouri. These employees took care of safeguarding the $650,000 in materials on hand and the sale of surplus materials that were in short supply.

When the war ended, KAMO was reactivated. Many of the poles already purchased were no longer usable. New plans called for increasing the line into Missouri to 66,000 volts. Delivery time for materials was now being measured in years instead of months.

The 1953 annual report states that in 1947, Empire Electric sent an offer to buy KAMO's facilities to REA. After meeting with KAMO's leaders, REA sent the utility a contract setting forth the conditions under which a sale could be made. Nothing was heard from Empire within the fifteen-day time frame set forth in the contract and as a result the proposal died.

Before the power could flow, this control room at the power plant had to be set up, no doubt a difficult task. Photo courtesy of KAMO Power.

Progress would be slow. However, at the time of the annual meeting in 1948, plans were well underway to provide power to fifteen electric cooperatives serving 42,162 members and with plans in the works to serve 62,757.

From the start, Irby Murray would represent Ozark Electric's interests on the KAMO board. KAMO had expenses of $550,000 due to shortages of materials, construction delays, and changes of plans resulting from wartime conditions and government production restrictions. Early plans called for just three cooperatives to be served at first. These included one Oklahoma system, Indian Electric Cooperative, along with Barry Electric and New-Mac Electric in Missouri. However, no one wanted these three systems to bear the entire start-up expense, given the planned expansion of KAMO.

In true cooperative spirit, Ozark Electric's board agreed to pay its portion of the KAMO start-up loan despite the fact that it would not receive power from the power cooperative for many years. The plan called for each member system to pay to KAMO within thirty days of the end of the calendar year its share of the debt owed by KAMO to REA, based on the number of kilowatt-hours purchased from its supplier. In return, KAMO would move ahead on construction as promptly as possible for New-Mac, Barry, and Indian Electric Cooperatives.

This agreement, also approved by the other systems that became members after the plan's approval, was critical to moving KAMO forward. Without it, the power cooperative had no income to pay its debt. Other Missouri systems joining included Southwest Electric, Barton County Electric, Sac-Osage Electric, and White River Valley Electric Cooperatives.

On April 26, 1946, Ozark's board approved a wholesale power contract with KAMO. The resolution read: "Ozark Electric has undertaken to purchase from KAMO Electric Cooperative all energy used by the member, when electric energy becomes available."

KAMO's vehicles, including this Chevy Suburban, were painted flat black. Photo courtesy of KAMO Power.

At the 1947 annual meeting, KAMO signed a power contract with the Grand River Dam Authority that would provide power at rates averaging less than ½¢ per kilowatt-hour. The board went on record favoring the sale of a steam plant owned by the War Assets Administration to the authority to firm up power output from the Grand River Dam located in Miami, Oklahoma.

"We believe it is best that this great power plant be permitted to produce electric energy for use of the people by distribution through a public agency (GRDA) at very low cost rather than be permitted to pass to private ownership for resale of electric energy at greatly increased costs," read a resolution sent to governors in four states, to Congress, and to the War Assets Administration.

A resolution passed at the January 1948 Ozark board meeting summed up the need for KAMO Power. This resolution asked REA to consider making loan funds available to either KAMO or Ozark for a transmission line to be built into the area from the presently proposed KAMO lines since the cooperative is "paying excessive wholesale rates and also because of inadequate power supply."

Let There Be Light!

Ozark Electric member C. Wendell Froshee submitted this humorous story for the book The Day the Lights Came On *about the time electricity first came to Rose Hill Baptist Church. The church is located five miles north of Willard on Highway Z.*

In 1941 when rural electric came to north Greene County, Rose Hill Baptist Church was one of the first places that had electricity. I was a boy of eight years and had never seen an electric light, so it was quite an event for our church and the neighborhood.

Here is what happened. We had the church wired and the church records state that we furnished the material and paid the man to wire it a total of ten dollars. After it was completed, the church thought it would be nice to have a little ceremony. So the first Wednesday night after it was done we all gathered for prayer meeting at the usual time.

The janitor had all the lamps trimmed and burning, at the proper time the oil lamps were put out and the switch turned on for the first time. Well, you guessed it, nothing happened. We were all in the dark, the lamps relit and a discussion followed.

Finally, one of our older men stood and addressed us and this is what he said: "We should not be alarmed because the electricity had to come all the way from Mount Vernon, Missouri." Being eight years old, that sounded good to me. However, some wiser minds checked things out and got it working.

Above: *A KAMO employee uses an A-frame hoist on one truck to unload equipment from the bed of another truck in 1958.* Photo courtesy of KAMO Power.

Left: *Meters in substations provided data for billing member systems for their wholesale power. By July 20, 1955, the last of Ozark Electric's substations located at Miller was transferred from Empire Electric to KAMO Power.* Photo courtesy of KAMO Power.

This massive truck and trailer, probably army surplus, hauled the heavy equipment used in substations. This photo is from 1956. Photo courtesy of KAMO Power.

KAMO's board made many trips to Washington, D.C., in order to secure financing for its lines. The board is shown here outside the offices of the Rural Electrification Administration. Ozark Electric's Irby Murray is shown in the center of the front row. He served as KAMO's vice president at the time. Photo courtesy of KAMO Power.

The situation was the same around the state. As electric cooperative membership increased dramatically, members were also discovering more ways to put electricity to work on the farm and in the home. As a result, cooperatives became second-class citizens when it came to power supply. At the mercy of their power supplier, they experienced rate increases and often saw their supply of current shut off at critical times when the supplier's native load had to come first. Leaders at Missouri's electric cooperatives believed a cooperative supply of power would end these shortages and also bring lower rates to members.

Not surprisingly, Empire Electric disagreed. The private power company, seeing its contracts for wholesale power with Ozark and other southwest Missouri electric cooperatives slipping away, opposed KAMO from the start. Empire filed a complaint with the Missouri Public Service Commission contending the cooperatives were making plans to expand without the commission's approval.

But KAMO answered the charge by saying it had complied with all Missouri laws and as a non-profit utility, it was not under the commission's jurisdiction. According to a story in the September 1948 *Rural Electric Missourian*, Empire also conducted what KAMO officials called a "windshield feasibility study" of the KAMO service area and came up with figures that showed the project to be unsound. REA officials were not impressed by the claims, recalling that power companies made similar claims in the early days of rural electrification, claiming the program was not needed and was doomed to failure.

Engineers compare notes at one of the early KAMO substations. This photo was taken in 1958. Photo courtesy of KAMO Power.

KAMO employees test the equipment at a new substation in the 1970s. Photo courtesy of KAMO Power.

The six generation and transmission cooperatives in Missouri would face constant opposition throughout their early years. In fact, a lawsuit filed by Kansas City Power & Light questioned the legality of contracts made with the Southwestern Power Administration. This legal maneuvering held up loan funds KAMO critically needed for many years until a federal court in Washington, D.C., upheld the contracts.

The Ozark board made its feelings on its relationship with KAMO known in a resolution passed on August 11, 1949. The resolution called for KAMO to set all power supply contracts and that regardless of the rates contracted for by KAMO with the various power sources, an equitable rate should be made to all member systems.

The tremendous growth on KAMO member systems, especially Ozark Electric, resulted in the need for additional transmission lines throughout the system. This line work took place in 1983. Photo courtesy of KAMO Power.

> It is believed that if this policy is once established and adhered to, it would wield KAMO into a truly compact organization which would not only strengthen KAMO but would command the respect of both public and private utilities. It is felt that nothing could engender more contention, hard feelings and criticism than a system of rate discrimination between KAMO's own membership. The strength of any cooperative organization has always been in the unity and loyalty of its membership. This has been definitely demonstrated by the history and the struggle of this cooperative from its inception right up to this very day.

REA showed its support for cooperative power when it approved a $10.5 million loan for KAMO in December 1950. The loan came after several trips to Washington, D.C., by Irby Murray and other directors. Murray, KAMO's vice president at the time, called the signing of the loan papers "one of the happiest occasions of his life."

As the power cooperative's lines began moving closer to Ozark Electric's substations, a representative from Empire Electric asked for a meeting with the board. He encouraged them "not to be hasty" in signing power contracts with KAMO, claiming Empire could continue meeting all of the cooperative's needs.

The work of building KAMO's transmission lines ran into problems not seen when Ozark Electric was formed. Perhaps not understanding the need for these lines, landowners in the Tulsa area did not want to grant right of way. In addition, a vandal destroyed one of the KAMO substations located at Cedar Springs.

KAMO energized its first substation in Missouri on May 20, 1949. This substation, located near Cassville, would serve Barry Electric Cooperative and would later be dedicated to KAMO's

A KAMO employee installs insulators at a new substation built in 1983. Photo courtesy of KAMO Power.

It takes nerves of steel to work on the tall structures that carry KAMO's transmission lines today. Photo courtesy of KAMO Power.

first president, "Uncle" John Hobbs, a farmer from Arkansas. That cooperative reported it would save $2,400 a month on its power bill through KAMO.

At the substation dedication, Missouri Attorney General J. E. Taylor answered charges of "communism" and "socialism" made by the private power companies who opposed their creation. He said he was against communism and socialism and any other "ism" except Americanism:

> I cannot think of any more democratic thing than giving electric service to our rural people as well as fellow city citizens. We know that electric service has made work lighter for the farmer and more comfort and lighter work for his wife. It has given the farms some of the comforts of the city. The time will come when every man, woman and child will have the benefits of electric power. This power comes to us as a result of natural resources which the people of this country should own and be benefited by. I have always been against and will always be against natural resources being controlled by a private utility.

By February 1954, work to tie in Ozark Electric's Verona, Clever, and Crane substations was underway, but was once again held back by shortages in material.

The long-awaited day when Ozark members would receive power from a cooperative supplier came on April 8, 1954, when the Monett substation was energized from KAMO's transmission lines. A few weeks later, the Clever and Crane substations followed.

On July 20, 1955, the Miller substation, the last substation served by Empire, was transferred to KAMO Power. Ozark Electric was now receiving all of its power from the generation and transmission cooperative it helped create. The previous month, Empire had offered an extension to the contract at a rate that was 1 mill higher than the KAMO rate, proof that the very existence of KAMO, even before it was delivering power, had kept wholesale rates artificially low.

A $4 million loan in 1959 would help KAMO meet the rapidly growing needs of its member systems. That loan was used to build 161 miles of 69,000-volt transmission lines, construct 18 miles of 115,000-volt line, make interconnections with other power cooperatives, and increase substation capacity.

Of this new construction, ninety-eight miles were in Missouri, including a line extending from Jenkins in Barry County to the new Table Rock Dam, which was under construction. This line offered an interconnection with the Southwestern Power Administration, giving KAMO systems access to other federally owned hydroelectric projects.

William "Arthur" Carrier was appointed to the Ozark Electric Board of Directors in December 1967, a position he held for more than thirty-eight years. A tireless crusader of the REA program, Mr. Carrier also served on the KAMO Power Board for thirty years and served as KAMO's representative on the Associated Electric Cooperative (AECI) Board for twelve years.

Another important step was an interconnection with Sho-Me Power at its substation in Buffalo, Missouri. This improved reliability for both power cooperatives.

Because of delays getting the transmission lines and substations built, KAMO never served electric cooperatives in Kansas and only a couple of Arkansas cooperatives stayed with KAMO. These systems turned to other G&Ts formed in their home states for wholesale power.

Over the years, KAMO has functioned as a "big brother" to Ozark Electric and its other member systems in Missouri and Oklahoma, providing them with ample electricity for their tremendous growth. To stay ahead of the demand for electricity, KAMO purchased 38 percent ownership in the Grand River Dam Authority's Unit 2 located at Chouteau, Oklahoma.

In 1997, KAMO officially became a member of the Associated Electric system. Associated, organized in 1961, functions as a generation cooperative, supplying electricity from coal, gas, hydro, and wind resources. Like the six other power cooperatives, KAMO maintains the transmission segment in a three-tier system that is unique to Missouri.

Given the tremendous obstacles placed in the way of KAMO's pioneers, it's a wonder the system ever got off the ground. However, perseverance was in large supply and every challenge was met with a solution. In the words of Irby Murray, who died August 27, 1961: "The KAMO system stands as a symbol of the initiative and resourcefulness of farm people in working together to solve their own power-supply problems."

In the 1950s, Ozark Electric had two of these spraying trucks working its many miles of right of way. Shown here are Billy Powell and Don Eskew.

Chapter 5
A Promise Fulfilled

By 1950, the promise of area coverage was nearly complete for Ozark Electric. In June of that year 12.4 miles of Section T were deleted because Empire Electric had already extended lines into the area.

Manager Claypool Daugherty reported it was now down to making short extensions of one half to one mile to make service available to all in the area who did not currently have it. The board agreed they should run service to anyone requesting it, even if the distance was more than a mile to serve one consumer.

One Stone County farmer appeared before the board to request service even though he was in the area served by White River Valley Electric Cooperative. The terrain there was so rough White River's lines could not reach him. Ozark Electric agreed to find a way to serve him. Another man requested service because the promised line from the City of Springfield had not materialized. In this case Ozark Electric worked with Southwestern Bell Telephone to extend service over its poles.

Ozark Electric's employees gather for the traditional Christmas party in 1954. The man standing is a young John Lomas.

Despite having power, many members were still drawing water from wells with buckets or hand pumps. According to an article in the April 1951 edition of the *Electric Sparks*, only half of Ozark Electric's members had electrified their water system. The newsletter pointed out that $5.40 would pump the 70,000 gallons of water a typical family of four would need in a year.

Marvin Carver, assistant manager, spoke to the board at the first meeting in 1951:

> This cooperative has expanded so rapidly that it is hard for some of us to realize the actual change that has been made. For the last several years we have had our thoughts on construction, for it was necessary to get as many members connected as soon as possible. Now the construction period is over and we are definitely in the operations and maintenance end of the business. Therefore, it is time for us to stand back and take a look at our operations and ourselves and set down as men and work out the most workable policies. We are in a big business and if we as key employees do not operate the business as a business, and I mean by that, a paying business, then the utilities will have a just cause to say that it is not business managed.

His words would seem prophetic in the summer of 1953, when REA demanded a proposed plan of future operation showing the cooperative's ability to repay its debt. Normally, REA allowed a five-year grace period before a cooperative had to start making payment on a loan. This allowed time for the line to be built and revenue to start coming in from those being served by it.

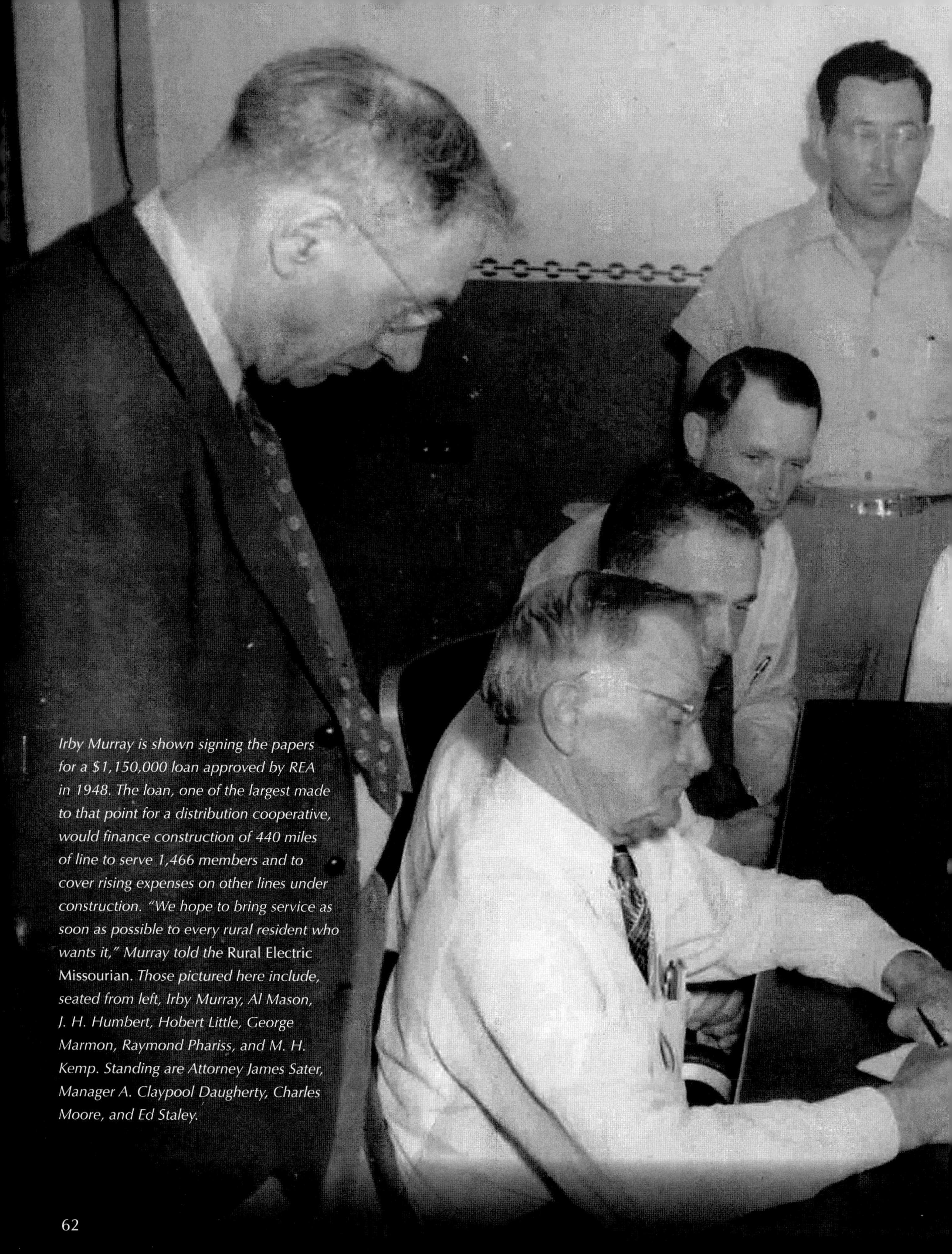

Irby Murray is shown signing the papers for a $1,150,000 loan approved by REA in 1948. The loan, one of the largest made to that point for a distribution cooperative, would finance construction of 440 miles of line to serve 1,466 members and to cover rising expenses on other lines under construction. "We hope to bring service as soon as possible to every rural resident who wants it," Murray told the Rural Electric Missourian. *Those pictured here include, seated from left, Irby Murray, Al Mason, J. H. Humbert, Hobert Little, George Marmon, Raymond Phariss, and M. H. Kemp. Standing are Attorney James Sater, Manager A. Claypool Daugherty, Charles Moore, and Ed Staley.*

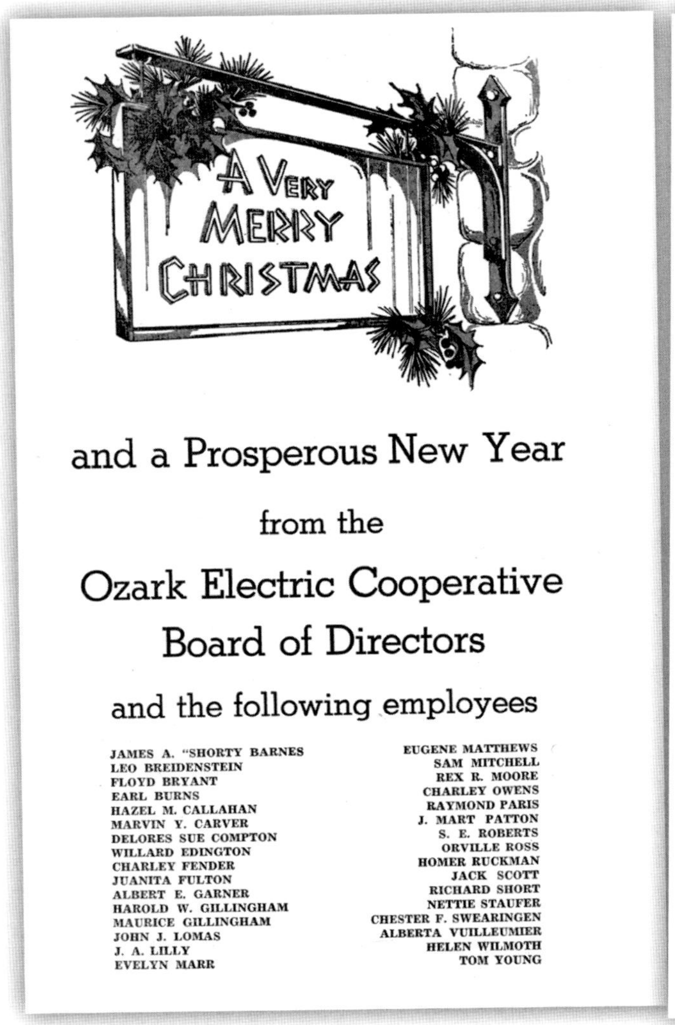

A Very Merry Christmas

and a Prosperous New Year

from the

Ozark Electric Cooperative

Board of Directors

and the following employees

JAMES A. "SHORTY BARNES
LEO BREIDENSTEIN
FLOYD BRYANT
EARL BURNS
HAZEL M. CALLAHAN
MARVIN Y. CARVER
DELORES SUE COMPTON
WILLARD EDINGTON
CHARLEY FENDER
JUANITA FULTON
ALBERT E. GARNER
HAROLD W. GILLINGHAM
MAURICE GILLINGHAM
JOHN J. LOMAS
J. A. LILLY
EVELYN MARR
EUGENE MATTHEWS
SAM MITCHELL
REX R. MOORE
CHARLEY OWENS
RAYMOND PARIS
J. MART PATTON
S. E. ROBERTS
ORVILLE ROSS
HOMER RUCKMAN
JACK SCOTT
RICHARD SHORT
NETTIE STAUFER
CHESTER F. SWEARINGEN
ALBERTA VUILLEUMIER
HELEN WILMOTH
TOM YOUNG

Ozark Electric had thirty-two employees when this 1954 Christmas ad ran in the newspaper.

Twentieth Annual Meeting

of the

Ozark Electric Cooperative

Mt. Vernon, Mo.

Wednesday, August 15, 1956

at the

High School Auditorium

Registration Starts
8:30 a. m.

Meeting Starts
10:00 a. m.

★ Slim Wilson

and his gang from KWTO and KYTV to entertain

★ Free Lunch

served at noon by Chamber of Commerce and "Jaycees" to Cooperative Members and their immediate families

Slim Wilson and his gang provided the entertainment when Ozark Electric held its annual meeting at the high school auditorium in August 1956.

But delays in getting materials—due to demand and the escalating Korean Conflict—meant construction was eating up most of this grace period. In some cases, Ozark Electric was paying for lines before it was receiving any income from them. Some members also had forgotten their promise to have their homes wired and ready to take electricity when it became available.

Those materials for building lines were costing more too. A mile of line cost $848 to build in 1939. By 1951, that cost had risen to $2,204. In 1951, rates were increased by placing a 10 percent surcharge on every bill. This was the first rate increase since 1939.

"We have just finished a year filled with many perplexing problems," Board President Irby Murray said in his annual meeting report. "We are living in a period of abnormal times, neither complete peace or complete war. This has forced upon us a state of uncertainty, which has made it very difficult to choose a course of action, or plan for the future operations of the cooperative."

Recognizing that Ozark Electric had grown faster than its revenues, the board took drastic steps to cut expenses and bring in more revenue. The Ash Grove office was closed, and all of the employees working there except one were laid off. The right-of-way department, power-use department, and engineering staff were eliminated. The newsletter was cancelled, retirement benefits ended, and memberships in related organizations were not renewed. In addition, several surplus vehicles were sold by sealed bid.

These were difficult times for the cooperative, which had worked hard to ensure all in its area could have the promised electricity. One of the problems was that while Ozark Electric had an abundance of members, many were using the minimum amount of electricity and were too poor to increase their usage, despite the cooperative's loans for wiring and appliances.

At its meeting in September 1953, Claypool Daugherty asked for permission to resign as manager. Assistant Manager Marvin Carver stepped in as interim manager until a new one could be hired.

There were twenty-three applicants for the manager position. However, the unanimous choice was a familiar face at the cooperative, S. E. Roberts. He returned to service at Ozark Electric in December 1953 and wasted no time in putting the cooperative back on its feet.

He told the board that Ozark Electric was too large for any one person to personally supervise all of its activities. For this reason, he reorganized the staff into four departments and appointed department heads including Marvin Carver, assistant manager; Chester Swearingen, operations and maintenance; E. C. Shafer, auditor and office manager; and Floyd Curnutt, power use. All of the department heads would now report at board meetings.

After extensive review of the new manager's budget and plans for the future, the board agreed it could afford to reopen the

Chester Swearingen was more at home working with the outside employees than in the front office, where he moved after becoming manager in 1964. He was hired in 1943 as a lineman but quickly worked his way up to superintendent of maintenance and operations.

Ozark Electric employees posed for this group shot in December 1956.

A-frames mounted to the back of trucks helped hoist the tall poles for crews building lines.

The poles got taller and there were many more of them in the 1950s as an all-out effort to provide electricity to all who wanted it got underway.

Ash Grove office in 1954. The cooperative's finances improved from a $21,631 deficit to $11,297 in the black in one year. In addition, Ozark Electric was able to place more than $250,000 in short-term investments in 1954. This led REA Field Representative A. J. Rowden to remark that he believed Ozark Electric Cooperative had made more progress within the past year than any other cooperative in the nation.

That year was the end of an era for a person who had helped guide Ozark Electric from its beginnings in 1937. Irby Murray announced at the February 1954 board meeting he was stepping down as a director due to his health and age. In response, the board passed the following resolution in honor of its president:

> WHEREAS, I.E. Murray, president and member of the Ozark Electric Cooperative, has this day tendered his resignation and the directors recognizing and appreciating his many years of service to the cause of rural electrification, adopt the following resolutions:
>
> BE IT THEREFORE RESOLVED that whereas I.E. Murray was one of the early pioneers in this section of the country in the cause of rural electrification and one of the few who had the vision that the life of farmers would really begin with the introduction of electricity, and through his efforts and a few others the Ozark Electric Cooperative was organized in 1937 and he became the president thereof, a position which he held until the present time; that during such a time the cause of rural electrification has had many problems and much opposition, all of which he met courageously and vigorously; and through

Crews use the A-frame truck to set a pole for the radio antenna.

Ozark Electric members gathered under the big top supplied by the statewide association for annual meetings beginning in the late 1940s. Ozark is one of the electric cooperatives still using the annual meeting tent.

> these many years despite opposition Ozark Electric Cooperative is one of the better and larger Cooperatives in the state. During these many years of service to the cause of rural electrification he made many sacrifices of personal nature and devoted his time, his energy and thoughts to the successful growth of the Ozark Electric Cooperative and the board regretfully accepts his resignation.
>
> IN RECOGNITION of his service and devotion to this organization, we confer upon him the honorable position "President Emeritus" and we sincerely say to you that as you retire from the activities you can rejoice in the fact that you have seen your vision of rural electrification become a reality, and that you have had a major part in such development, and that this board and its members may all well say "Well done thou good and faithful servant."

He recalled he could have had private power company electricity on his farm near Ash Grove, but the "subsidy" they asked for not only went against the grain, "but we couldn't quite afford it. There is no better round-the-clock helper around the farm than rural cooperative electricity that the farmer can afford."

Few people embraced the concept of rural electrification better than Irby, who used the newfound electricity to better preserve fruit from his extensive orchards in an electrically cooled barn. He also made sure his wife was well supplied with the latest electric appliances and

As construction foreman, Charlie Fender was responsible for coordinating the many power lines built in the 1950s.

Evelyn Marr proved to be one of Ozark Electric's longest-term employees, working forty-five years as a materials clerk.

encouraged members to "go electric" whenever possible.

Cattleman Ed Staley of Willard was elected as the new president, a post he would hold well into the future.

At this time the City of Lockwood asked Ozark Electric to take over supplying electricity to the city. Initially, REA indicated it would loan the cooperative the $33,000 it would need to build the line and a new substation in order to serve the town's eight hundred residents. That loan would not be approved, because Lockwood residents already had electricity, but REA did okay letting the work be done using general funds. Lockwood began receiving power from Ozark Electric on October 21, 1954.

As the REA program turned twenty in 1955, a story in the *Lawrence Chieftain* looked back on progress made at Ozark Electric. When the first lines were energized, members used an average of 38.8 kilowatt-hours per month. Every five years the membership and its electricity use had doubled. In 1955, there were 7,552 members, including 201 schools and churches and 122 commercial accounts. They were using an average of 239 kilowatt-hours per month.

The cooperative now had 2,800 miles of line and thirty-seven employees. Its eighteen vehicles were driven more than 25,000 miles per month. Payroll, it was noted, pumped between $10,000 and $11,000 a month into the local economy.

Now that construction had slowed, Ozark Electric's manager put public relations on the front burner. Earl Burns, who had been working in the right-of-way and staking departments, moved to a new role as the cooperative's public relations man. His first duty was to restart the cooperative's newsletter, called *The Electric*

Sparks. A new logo and a special paint scheme were added to all vehicles.

In another nod to public relations, Ozark Electric's employees were paid with silver dollars in the spring of 1956. As the dollars rippled through the communities, business people could see the impact Ozark Electric was making on the economy. A year later this idea was repeated using $2 bills.

At this time the annual report of REA showed only one REA-funded system in Missouri, M&A Power, had greater margins than Ozark Electric in 1955. The minutes show Ozark had paid REA $707,009.21 in principal and $617,964.88 in interest on its loans. It had cash investments of $817,520. It had made advance payments to REA of $108,457.24. Equity in the system had increased to 13 percent.

A year later, the cooperative had more than a million dollars invested in bonds and advance payments to REA. Clearly, the cooperative again was in excellent financial shape. This allowed Ozark Electric to absorb a 27 percent rate increase on wholesale power passed down from the Southwestern Power Administration to KAMO without raising rates.

Another big project for the cooperative was the relocation of lines on the southern end of the system due to the construction of Table Rock Lake. An additional ten miles of line were added to the system to serve the anticipated growth in resorts and other commercial operations at the lake. Over the years, this area would be the source of tremendous growth for the cooperative.

In February 1957, representatives from Barry Electric, Ozark Electric, White River Valley Electric, and Carroll Electric based in Berryville, Arkansas, met to set boundaries in the Table Rock Lake area. After agreeing who

This crew worked out of a branch office in Ash Grove during the 1950s.

A squirrel scampers down a pole as Bud Law prepares for action in the 1950s. Those who remember Bud were honored to work with him. He helped train many of Ozark Electric's linemen.

Charlie Fender checks the equipment on this Chevrolet truck.

Director Raymond Phariss, shown here setting up an electric heater in his dairy, was one of the early pioneers who helped convince reluctant members to join the cooperative.

would serve where, White River's manager suggested his cooperative sell 7.5 miles of line and twenty-three members located on Aunt's Creek Road to Ozark Electric to avoid duplication of service and extensive travel since one of the roads used to service these accounts would be flooded when the lake filled.

Mother Nature played havoc with the cooperative's lines in May 1957 when a powerful storm swept through the area, killing power on 1,500 miles of line. The damage required the replacement of eight poles and ten transformers, repair of seventy-five separate lines, and three thousand miles in travel. This storm prompted the board to set up a fund equaling $100 per mile of line as a reserve against damage from ice and windstorms.

In 1958 Ozark Electric started its tradition of awarding winners in the FFA Chapter Project Contest. The award was an electric drill, given to the person in each chapter who came up with the best laborsaving electrical device. For fifty-four years, Ozark Electric has carried on this tradition of awarding an electric drill to area FFA chapter members.

Farm use of electricity was certainly on the rise, perhaps promoted by the FFA youth. One Ozark member touted the advantages of electrically powered irrigation in the August 1953 *Rural Electric Missourian*. W. D. Owens, who raised tomatoes west of Mount Vernon, said, "I would consider it a wise move for any farmer in the area to mortgage his farm if that were necessary to install irrigation. That loan would be more secure than any investment I can think of, because a crop is more nearly guaranteed by irrigation than is possible any other way."

As the decade drew to a close, REA reported Ozark Electric was first in the state in farm usage with a monthly average of 288 kilowatt-hours per farm. Apparently the efforts of the power-use department were paying dividends. A board committee came up with a twelve-point power-use plan aimed at increasing sales. This plan included close work with youth programs such as 4-H and FFA.

Ozark Electric would also join other electric cooperatives in southwest Missouri in a TV program on KYTV in Springfield. Sho-Me Power organized the participation in the *High Noon* TV program. The show ran from 12 to 12:30 p.m. daily from Monday through Friday.

A decade-ending statistical report from REA showed Ozark Electric with the second highest net margin for any electric cooperative in the state, behind only Ozark Border Electric based in Poplar Bluff. The minutes, perhaps suggesting a few sour grapes were swallowed by the second-place status, reported Ozark Border was ahead due to it serving six thousand more members, "eight or 10 towns," and from personal observation, "their preventative maintenance program is inadequate."

Whatever its ranking among the state's electric cooperatives, Ozark Electric would end the decade in excellent shape, ready for the challenges ahead.

Jim Armstrong, Byford White, and Maurice Gillingham pose in front of one of the co-op's trucks in 1947.

Two of Ozark Electric's linemen prepare to start their day. Members with phones could dial 122 to reach the cooperative.

The cooperative's board took the stage at the 1966 annual meeting held at the Mount Vernon High School auditorium.

Chapter 6
The Golden Age for Electricity

The 1960s were a golden time for Ozark Electric and electric cooperatives nationwide. They had grown exponentially and were celebrating their own twenty-fifth anniversaries along with the silver anniversary of REA itself.

Closer to home, Ozark Electric was in excellent shape. With KAMO's construction projects in Missouri complete, the cooperative now had a source for low-cost, reliable power. No longer would it have to worry about growth outpacing the supply of electricity; nor wholesale power rate increases that were out of its control.

For the first time, houses with no chimneys were being built as members went "all electric" in their homes. The July 1964 edition of the *Electric Sparks* touted the advantage of the "Flameless Home" owned by the Ralph Simmons family of Greenfield. It also told about the modern dairy owned by Melvin Backs of Lockwood, who installed his first milking machine in 1952. His Grade A dairy could now milk three cows at once and included a waterline for disinfecting udders.

The first retiree for Ozark Electric was Eugene Matthews, left, being congratulated by Manager S. E. Roberts in 1962.

Members installed the latest in modern conveniences, including electrically operated garage door openers, door bells, air conditioning, and electric heat. The latter signaled the end of cutting firewood or putting up with the soot and smell of fuel oil or coal furnaces.

Air conditioning, a modern marvel invented in 1902 by Willis Carrier, was slow to catch on in rural areas but began widespread use in the early 1960s. In 1963, members could experience this summer wonder at the newly air-conditioned Ozark Electric office in Mount Vernon.

Ozark Electric's power-use department helped members make the most of their new electric appliances. Ozark Electric invested $48,000 in upgrading its lines in Stone County to meet the growing demand for electricity. Heavy growth in the Springfield area also required considerable resources. While member connections had slowed considerably from the 1950s, an REA loan in 1960 brought service to an additional five hundred families. By 1964, average monthly electricity use by members had again doubled, this time to 414 kilowatt-hours per month.

Ozark's crews were kept busy building lines to new industries moving into the area. One was the McQuerter Chemical Company, a fertilizer plant built outside Mount Vernon. Growth was so heavy in the Table Rock Lake area, as expected, that KAMO began work on a new transmission line into the area. KAMO agreed to build a line from Jenkins to Cape Fair, connecting with Table Rock Dam. This line would also continue on to supply substations for White River Valley Electric.

A year later, a system study showed the Clever feeder line soon would not have the capacity to supply the demand in the area. A resolution asked KAMO to build a new substation at Battlefield and have it in operation as soon as possible.

Another potential major load was the baseball school opened by Mickey Owen, a major-league catcher who played for the St. Louis Cardinals, Brooklyn Dodgers, Chicago Cubs, and Boston Red Sox. Owen, who grew up in Nixa, opened the school on Route 66 near Miller in 1959 and ran it until 1963 when he sold it. The school remained open for fifty years and is still used for

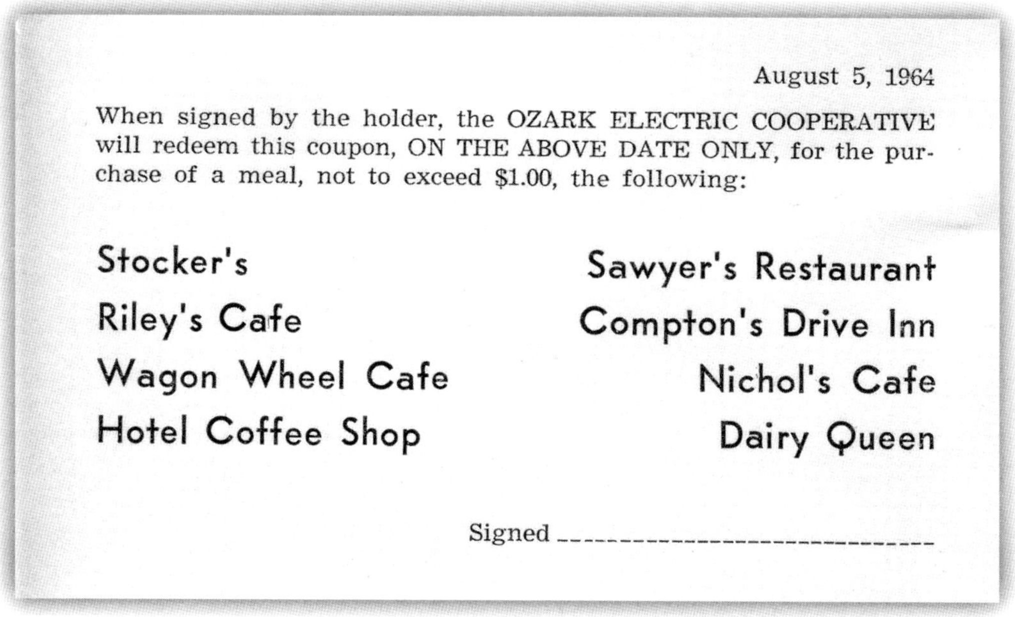

August 5, 1964

When signed by the holder, the OZARK ELECTRIC COOPERATIVE will redeem this coupon, ON THE ABOVE DATE ONLY, for the purchase of a meal, not to exceed $1.00, the following:

Stocker's	Sawyer's Restaurant
Riley's Cafe	Compton's Drive Inn
Wagon Wheel Cafe	Nichol's Cafe
Hotel Coffee Shop	Dairy Queen

Signed ______________________________

Those who attended the 1964 annual meeting received one of these coupons good for a free lunch at several restaurants in Mount Vernon. The cooperative stimulated the local economy in many ways.

tournaments. Among its alumni are NBA star Michael Jordan, actor Charlie Sheen, and Yankees Manager Joe Girardi. While the board approved building a three-phase line to serve the school, there's no record of it ever connecting to Ozark's lines.

Yet another project that kept the cooperative busy was the construction of Interstate 44, which required moving lines out of the right of way. While some portions of I-44 in Missouri were just re-signed sections of Route 66, in the Mount Vernon area an entirely new highway was built, requiring extensive relocation of power lines between Mount Vernon and Halltown and elsewhere.

Finances were so good that by 1961, Ozark Electric had made more than $1 million in advance payments to REA, tops in the state. Ozark also passed Ozark Border in net margins earned for 1960.

This allowed the board to begin paying capital credits to members. In a cooperative, any margins above and beyond the cost of doing business are retained for a period of time as the member's investment in the business. Then these margins are refunded based on the amount of business the member did with the cooperative.

In contrast, an investor-owned business receives money from the sale of stock to investors, who may or may not use the company's services. Any profits go back to the investors, not the people using its service.

In March 1963, a new capital credit department with Betty Richardson in charge began operation. Checks totaling $222,906.31 were sent to members. Ozark Electric's capital credit payments included the City of Crane, which paid Ozark and KAMO to wheel electricity across its transmission lines for resale to residents.

Big changes came for the power supply situation on January 13, 1961, when Associated Electric Cooperative was formed. Referred to as a "Super Cooperative" in Ozark Electric's minutes, Associated would, in time, take over generation duties for the six regional power cooperatives.

Ice cream was a big hit at the 1966 annual meeting.

This group of girls hold fans given out at the 1966 annual meeting. Shown with them is board member Ed Staley.

The new cooperative would allow member systems to control power from Bull Shoals and Table Rock Lakes and was expected to reduce wholesale power bills. KAMO and the other five regional power cooperatives would remain to maintain the high-voltage transmission lines that brought power to the distribution cooperatives such as Ozark Electric. This three-tier system—generation, transmission, and distribution—is unique to Missouri and serves the state's electric cooperatives well.

In 1966, Associated would complete its first coal-fired power plant at Thomas Hill in north-central Missouri. When this plant came on line, it let the state's electric cooperatives control their own destiny during a time when use of electricity was skyrocketing. Today, Associated Electric owns coal-fired power plants at Thomas Hill and New Madrid, along with natural gas plants in Glennonville, Missouri; Dell, Arkansas; and Chouteau, Oklahoma. It also buys the entire output of four wind farms located in northwest Missouri and Oklahoma.

Employees and directors of Ozark Electric were saddened by the death of long-time manager S. E. Roberts on June 1, 1964, at age sixty-eight. Roberts was hired at Ozark Electric on January 7, 1939, as superintendent of maintenance. He later served as assistant manager and took over the top spot on October 22, 1940. Except for a ten-year term when he was manager of Sho-Me Power, he had remained at the helm of Ozark Electric through many challenges.

Ozark Electric sported a bowling team that included Manager Chester Swearingen, left, in the 1960s.

Earl Burns and Marvin Carver stand in front of the office in the early 1960s.

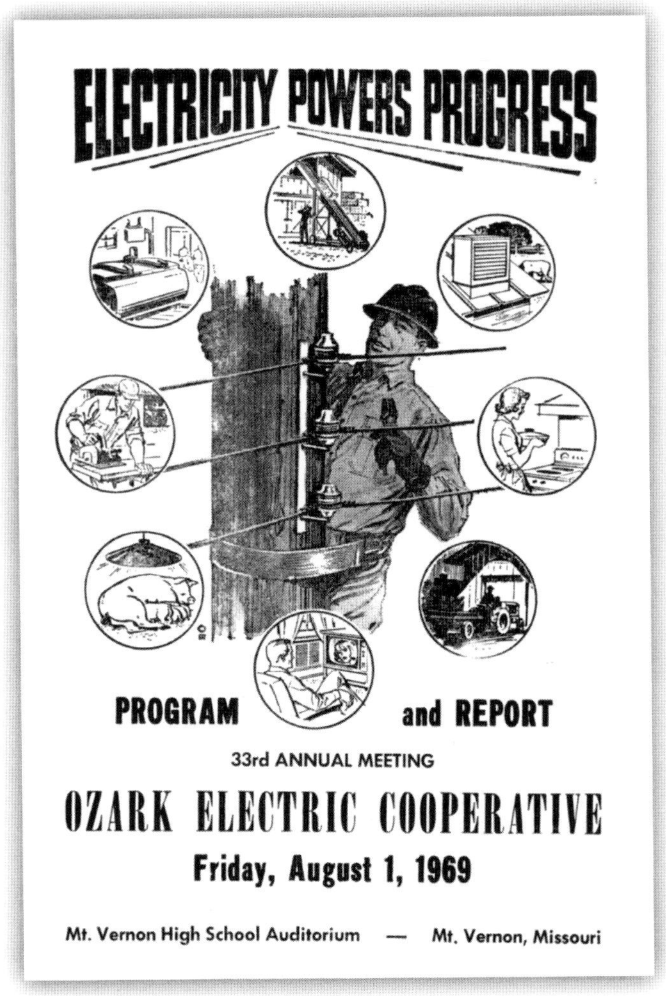

This was the program for the 1969 annual meeting.

Mrs. Joe Williams told how much she loved cooking on her electric range in an issue of the Electric Sparks, *Ozark Electric's newsletter.*

His successor was no stranger to the cooperative. Chester Swearingen was hired as a lineman in 1943 when the cooperative lost one of its two maintenance linemen to the war effort. At first he was offered a full-time job repairing the co-ops' trucks, a job he was doing part-time from a garage he operated in downtown Mount Vernon. But since he already had experience as a lineman for Kansas City Power & Light, he was hired to fill that role. While still in high school at Mount Vernon and later Miller, he had done electrical contracting in his own business. He also operated the Miller power plant.

He was later promoted to foreman, then superintendent of maintenance and operations. In this role, he led Ozark Electric's crews through its years of rapid growth and oversaw the construction of powerlines until the point where two letters were needed to identify them. He would serve as manager until 1974.

According to his son Jerry, all of the cars and trucks Chester drove while employed at Ozark Electric were No. 13. He said that gave him good luck, which it must have considering all of the safe driving pins he was awarded. He never had an accident during his career.

The job of an Ozark lineman in the 1960s was hard and dangerous work. Most outside employees learned on the job along with safety lessons taught by instructors from the Missouri State Rural Electrification Association who traveled around the state putting on safety meetings. Don Wood, who began working at Ozark in 1962, recalls hard hats becoming mandatory not long after he started. Ozark's men wore orange ones, he said.

Crews used air-powered jackhammers to bust through the limestone when digging holes for poles. Sometimes they resorted to unusual tactics. Don remembers one job where his

Earl Burns visits with Ozark Electric Board President Ed Staley outside his barn, a landmark on his farm near Willard.

Cattleman Ed Staley took over as board president when Irby Murray retired.

In the 1960s, reliance on electricity meant linemen had to work while the lines were hot to avoid service interruptions. Linemen used hot sticks to work on the lines. The hot sticks were carried in a special trailer to keep them safe.

Young lineman Jack Swearingen (right), shown here with lineman Bud Law, looking at a weather balloon parachute they found in the power lines.

crew struggled to set a pole in a swampy area near Bois D' Arc. Every time they tried to dig the hole, the sides collapsed filling it back in. "We had heard a story that you could tie a stick of dynamite to the end of the pole," he said. "When you set it off, the pole would fall right in the hole. We tried it and it worked."

John Lomas, who started working for Ozark in 1954, got his training digging holes for contractor Ray Foree when he built a three-phase line near Lockwood. "I didn't know a thing about electricity when I started working. It was all on-the-job training."

John remembers dynamite being extensively used in those days due to the rocky terrain. One of his teachers was a man named Wilkie who demonstrated how to set a charge for a pole hole. When the charge was set off, it blew everything out of a window in a nearby farmhouse. "The next time he loaded up, he said, 'I didn't break anything.' There was nothing left to break!"

In this era, crews began working lines hot, first using hot sticks that were carried in a special trailer. "Most linemen weren't interested in working live line work," Don recalls. "We worked hot, but we didn't use rubber gloves. We used hot sticks."

Adds John, "All of our training was to stay away from the hot lines."

Jack Swearingen, another former Ozark lineman, says live line work was done slowly and carefully in those days before bucket trucks and rubber gloves made the job easier. "If we changed out a pole hot on a three-phase line, we got one done a day."

These three linemen reminisced about their time at Ozark Electric. They recalled working as many as eighty-eight hours in one week while restoring power in a major winter storm. They used helicopters to patrol lines when roads were impassable, thanks in part to Jack's connections with the Missouri National Guard, which provided the choppers.

The work was hard, but the antics of their fellow employees kept them from getting discouraged. On a slow day, for example, John would get on the office PA system and announce, "There will be no work in the pineapple field today," an announcement he heard often when he was in Hawaii.

The linemen never knew what they would run into when restoring power. "When you work trouble calls you would think you've seen everything, but then find out you haven't," Jack said. "The most unusual thing I remember causing an outage was a frog. It was so small, you couldn't see it."

One of the biggest fears for a lineman was burning a pole. Linemen in the early days did all of their work from the poles, which they climbed using sharpened spikes attached to their boots. If the spikes kicked out, the lineman would slide down the creosote-impregnated pole, burning his arms and legs.

"My first day at work with John, he commented 'I never burned a pole, but a few have burned me.' The next pole I went up I burned," recalls Jack.

Pole climbing is still done despite the common use of bucket trucks today. In fact, surviving climbing school is mandatory for new linemen wanting to learn the trade.

Linemen in those days did a little bit of everything, including reading meters and disconnecting members who did not pay their bill, a task they all hated. Don remembers one young lineman who thought he should only have to climb poles. Storming into the line superintendent's office after a day spent reading meters, he threw his gloves on the desk and said, "If that's all you've got for me to do, I'm done."

For several years, the Ozark Electric Cooperative Annual Meeting of the membership was held in the Mt. Vernon High School Auditorium as shown here on August 5, 1966.

Instructors from the Missouri State Rural Electrification Association, now called the Association of Missouri Electric Cooperatives, provided safety training for Ozark Electric's crews in the 1960s. The linemen were learning pole-top rescue when this photo was taken.

Ozark Electric supported the construction of Table Rock Dam and benefited from its completion. Hydropower generated at the project is still used by the members, and tourism greatly benefits local businesses. The area around the dam became one of the chief sources of growth for the cooperative as well.

One of Ozark Electric's key accounts is Fantastic Caverns located north of Springfield. The show cave, discovered by a dog in 1862, offers the nation's only ride-through cave experience. This group took the tour in 1969.

"Well, I guess you are done," Don remembers him being told. "Maintenance crews did what needed to be done."

Ozark Electric continued its support of youth programs in the 1960s by taking part in the Youth Tour to Washington, D.C. The Youth Tour came from a challenge issued by then-Senator Lyndon Johnson, who encouraged the electric cooperatives to send young people to the nation's capital in order to create future leaders.

In 1965, Ozark Electric sent Charlotte Ann Burns and Greg Lynn Curl, both of Nixa, on the Youth Tour after the two were winners in the contest. The success of the contest prompted Ozark Electric Assistant Manager Marvin Carver to say the cooperative should continue its participation because "the cooperative story was told in many homes where it had not been discussed before, through not only the winners, but all others who participated in the program."

Ozark would take part in the Youth Tour until 1988, when participation was discontinued for ten years. In 1998 it was resumed and continues today, with several Ozark Electric delegates being selected for the prestigious Youth Leadership Council. Ozark's 1985 Youth Tour delegate, David McNabb from Republic High School, currently serves as Associated Electric's chief financial officer.

In the closing years of the 1960s, Ozark Electric had reached 9,529 members who were using an average of 582 kilowatt-hours per month, paying an average of $13.55. There were now 3,077 miles of line and 59,749 poles. Service now required nine substations and the number of employees had increased to forty-seven.

Willie Wiredhand:
A Spokes-Plug for All Time

Willie Wiredhand, the little guy with a light-socket head, push-button nose, and an electrical plug for his bottom and legs, celebrated fifty years as co-op spokes-character and mascot in 2001. He's appeared in just about every type of cooperative publication and promotional item: signage, stationery, newsletters, annual reports, advertising, Christmas ornaments, coffee mugs, watches, shirts, baseball jerseys, beach towels, night lights, bobbleheads, and more.

Willie's actual birthdate is traced to October 30, 1950. He was the creation of Andrew McLay, a freelance artist for the National Rural Electric Cooperative Association in Washington, D.C.

"We were toying with ideas for a rural electrification symbol," recalled William S. Roberts in a tribute to McLay, who died of cancer at age fifty-two in 1974. Roberts was editor of *Rural Electrification* magazine, NRECA's trade publication, in the 1950s. "I had tossed out the idea that the symbol ought somehow to portray rural electric service as the farmer's hired hand."

Sprawled out on Roberts's living room floor, McLay gave birth to "Willie the Wired Hand." NRECA's membership selected Willie as their animated ambassador at their national meeting in February 1951. Willie's name was soon shortened to "Willie Wiredhand."

Willie came along in the heyday of animated advertising characters. Hundreds, promoting everything from bleach to antacids, were already in the marketplace. For whatever reason, people connected and responded to these characters. They touched on a human need to personify things.

Though most of rural America had power by 1949, many consumers were still in the dark when it came to understanding electricity and the ways it could improve the farm and the lives of farm families. Willie, posed in front of everything from fans to ironing boards, explained the many ways electricity could help around the home.

Though Willie symbolized co-op friendliness, he also embodied co-op spunk, willing to stand up for what was right in the face of impossible odds. "He's small, but he's wirey" became part of Willie's trademark, which was granted by the U.S. Patent Office in 1957. The registration allowed Willie to become the beloved character he remains today.

Though he's no longer used as much as he was before, Willie's place is secure in the hearts of many consumers. He is truly a "spokes-plug" for all time.

Ozark Electric again held its annual meeting under the big top tent in the 1970s, a practice that continues today.

Chapter 7
A Crisis for Energy

The 1970s began with a word that would become a standard for Ozark Electric to the present time: growth. In his manager's column for the January 1970 *Electric Sparks*, Chester Swearingen predicted the new decade would be a busy one for Ozark Electric's crews. The 1970 work plan called for an additional eighteen miles of line to serve new members, as well as a new substation for the Lockwood area.

Assistant Manager Marvin Carver also talked about growth in his column. He noted there were eighty-four additional work orders in 1969 compared to 1968. Not all of these were for new service. Many were to increase the capacity of lines overburdened by new uses for electricity. In 1969, Ozark Electric ordered five hundred new larger transformers.

A U.S. Department of Agriculture report showed farmers were using an average of 10,536 kilowatt-hours, an increase of 720 kilowatt-hours from the previous year. That was the largest

Employees gathered at the Lion's Club in May 1972 to be recognized for working safely.

increase since 1959, and the load was expected to double in the next ten years. This meant the cooperative would have to continue its efforts to increase capacity with heavier lines, more substations, and larger transformers.

A year later, Carver was reporting "growing pains" as the cooperative struggled to keep ahead of new services, particularly in the southern end of the system and around Springfield in the Battlefield area. Even with the addition of a contractor's crew, demand from new homes and businesses outpaced the efforts of the workers. Added to the workload was the need to change out many of the aging poles. In 1970, more than 1,400 were replaced.

An estimate load study conducted for the cooperative showed the 84 million kilowatt-hours purchased in 1970 would increase to 129 million in 1975 and 200 million by 1980. "One can readily see that to design and build a system that will be required to deliver 200 million kilowatt-hours to the members by 1980, which is more than double the 1970 usage, will require, within the next 10 years, almost as much capital investment as has been required in the past 33 years," Carver wrote in the January 1971 *Electric Sparks*.

In 1971, Ozark Electric's membership topped ten thousand for the first time. Two years later, it had passed eleven thousand and showed no signs of stopping. The growth was coming from two directions. One was people escaping the congestion of Springfield—Missouri's third largest city. The other was people retiring or building vacation homes around Table Rock Lake.

All of this growth was outpacing the loans from REA, which hadn't approved a loan for Ozark Electric since 1969. Instead, cash reserves were used to build and improve service. Trouble came when REA began to increase its interest rate on loans. In the early days, REA loaned money as cheap as 2 percent.

But no one in the late '60s and early '70s was taking cheap interest for granted. Just like they did with power plants, Ozark Electric joined with other electric cooperatives to form their own bank in 1969 called the National Rural Electric Cooperative Finance Corporation, or CFC for short. If REA loans did become scarce, the cooperative still had a way to get the critically needed financing, although everyone knew the rates would be higher.

Ozark Electric wasn't the only Missouri electric cooperative seeing demand for power on the rise. It was obvious demand would soon outpace the available generation from Associated Electric, which now had two coal-fired units at Thomas Hill along with peaking power from the dams in Missouri and Arkansas.

Soon plans for a new power plant located in New Madrid, Missouri, were in the works. Each electric cooperative would have to pay its share of the new power plant, which was expected to cost $240 million. This would be the first rung in a ladder leading to higher rates for members.

In 1974, most of the original office building was demolished to make way for a larger structure.

"Our projections indicate that, by the time the plant is completed, in the next four to five years, we will be urgently needing it to supply the cooperatives of Missouri with the power that they need," Ozark President Ed Staley told the crowd at the 1971 annual meeting. "As all these costs climb, there can only be one inevitable result: costs will exceed income, and we will be forced to increase rates to continue to operate."

Gerry Diddle, manager of Associated Electric Cooperative, explaining a wholesale rate increase in August 1974, had this to say: "We have reached the end of the use-more-power-to-get-a-cheaper-rate era. Our system load for the entire state has doubled in six years. Just think: In the next six years we need to pick up a load equivalent to that developed in the first 38 years of the Missouri REC program."

Two big changes came to the cooperative in 1974. One of these was construction on a new office building. The original office, built in 1941, was torn down after the inside staff moved into a trailer. Construction was completed in the fall, giving the employees enough room to take care of the 11,600 members' needs.

The other was the retirement of Chester Swearingen after thirty-one years of service. His replacement was Tommy Sutton, who came to Ozark from Petit Jean Electric Cooperative in Arkansas.

As the new manager took over, the cooperative's crews could not keep up with the requests for new connects. Crews began working ten-hour days, six days a week, in a valiant attempt to keep pace with new home construction.

Members learned of a new computerized billing procedure in the March 1974 *Electric Sparks*. No bills were due in April. Instead,

The foundation is being laid in this photo of construction on the new office building in 1974.

The new office building completed in 1974 is still in use today.

A broken pole, no doubt hit by a car, needs replaced. Ozark's trucks at this time were yellow.

members read their meters and sent the reading into the office, where staff members entered the information into a computer and printed out a bill.

Now a new threat to cooperatives nationwide caused considerable concern at the board table.

In 1968 Richard Nixon had been elected president. The new president proved hostile to the rural electrification program. In 1973, with a stroke of his pen, he issued an executive order abolishing the REA loan program, declaring that 80 percent of the loans were going for "country clubs and dilettantes… and others who can afford living in the country." What followed was an incredible grassroots rally held in the Mayflower Hotel in Washington, D.C. Rural electric leaders from across the country gathered under a huge banner that read, "We Protest." In the end electric cooperatives had more friends in Congress than they did in the White House and REA reemerged, but with interest rates now at 5 percent.

As if a new power plant, higher interest rates, a president who was hostile to the REA program, and increased costs for materials weren't enough pressure on electric bills, another global event would be the final straw forcing rates to skyrocket.

In 1973 Arab oil-producing nations started the first energy crisis when they put an embargo on oil in retaliation for U.S. support of Israel in the Yom Kippur War. The second crisis for energy came in 1979 in the wake of the Iranian Revolution. For the first time Americans had to deal with shortages of energy, particularly gas for transportation. Ozark Electric promoted conservation with this slogan: "Electrify: Conserve Fuels in Short Supply." President Nixon asked Americans to curtail their use of Christmas lights. Later, President Carter would don a sweater on national television and ask Americans to lower their thermostats in winter.

New Madrid Unit 2 came on line in June of 1977. But a year earlier, projections showed the phenomenal load growth would require more generation by 1981 or 1982. So once again plans were drawn up for a new generating unit even larger than the New Madrid unit. This one would be dubbed Thomas Hill Unit 3. It would be sited near Moberly in north-central Missouri where the first two of Associated's generators were built.

To build this plant Associated negotiated the largest loan guarantee in the history of the rural electrification program—a whopping $1.4 billion. The Thomas Hill Power Plant cost would be the "straw that broke the camel's back," forcing Ozark Electric's board to reluctantly put sizeable rate increases into effect. Though the plant was expensive, it would later be the deciding factor in a lengthy period of stable rates that would last well into the 1990s.

In 1973, the first rate increase since the 1930s was put into place. That one was small in comparison to the 20 percent increase the board reluctantly put into effect in 1975. Staff assistant Larry Austin, who would later become manager of Crawford Electric Cooperative, explained the 20 percent rate increase in the September 1975 *Electric Sparks*.

He said since the 1973 rate increase, wholesale power costs had increased 82 percent, largely due to increases in the price of coal. In one month the wholesale power bill had increased by $27,691. The price of a 35-foot pole had gone up 82 percent, transformers 34 percent, meters 15 percent, and fuel for trucks 10 percent. Loan rates

Despite advances in equipment, setting poles continued to be hard work when this photo was taken in 1976.

Key staff members with Manager Tommy Sutton in 1976. Clockwise from left are Lee Coleman, Don Wood, Jerry Mayberry, Sutton, Larry Austin, and Richard Neely. Austin went on to become manager of Crawford Electric Cooperative.

Transformers are loaded on a truck during a day on the job in July 1977.

had gone up 150 percent, not including the supplemental loans from CFC that were costing 9.25 percent interest.

Employees bid farewell to two long-time employees in 1973, Orville Ross and Earl Burns. Orville worked for Ozark Electric for thirty-one years. One of his first jobs in 1944 was to document transformers and other equipment installed in the field, since this was done without any record keeping. He climbed fifty to seventy-five poles a day compiling an accurate record of installations until a fall from an ice-covered pole forced him to abandon his climbing hooks and move inside.

Earl was in many ways the face of Ozark Electric. Though he started his cooperative career in the engineering department, he quickly became the power-use advisor. It was his job to increase the use of electricity by advising members on new appliances and tools to make life easier. He also took charge of the youth programs, working with 4-H and FFA members in the area.

Wholesale power costs continued to climb in 1976, when a 35 percent rate increase was passed down from Associated to KAMO to Ozark. The chief cause for this was a dramatic increase in the price for coal used to generate electricity. A graph in the March 1976 *Electric Sparks* showed the price for a ton of coal had increased from $4.95 in 1970 to $10.60 in 1975.

Still, the price of electricity for Ozark members compared favorably to rates paid by urban dwellers. Ozark members paid 2.57¢ per kilowatt-hour compared to 3.5¢ in St. Louis and 9.10¢ for New York City.

On January 1, 1977, the Ash Grove office was again closed as a cost-cutting measure. This move proved controversial, and it led to the resignation of Ozark Electric Board

Juanita Fulton works with one of the cooperative's early computers in 1977.

Ozark's engineering department in 1977. Shown here are Lee Coleman, left, and Raymond Paris.

Ozark Electric Cooperative Board of Directors in 1978. Front row, from left: Jess Williams, Jay King, Manager Tommy Sutton, Don Carlock, and Jimmie Crouch. Back row: Arthur Carrier, James Cox, Attorney Almon Maus, Doug Stone, David Trogdon, and Robert Neal.

Pat Prewitt, left, and Jerry Mayberry hand out humorous gifts during a staff Christmas party in the 1970s.

President Ed Staley and board member Max King, who objected to the closure.

Arthur Carrier, a cattleman from Lockwood, became the new board president, a position he would hold for many years. He came on the Ozark Electric Cooperative Board in 1967. Arthur also served on KAMO's board and as KAMO's representative to the Associated Electric board.

More bad news for power generation came on December 6, 1977, when a nationwide coal strike began. At this time, Associated had a sixty-day supply of coal for its power plants. In order to stretch that supply out over the lengthy strike, expensive electricity had to be purchased from utilities that generated with fuel oil or natural gas. The costs trickled down to the distribution cooperatives such as Ozark and ultimately to the end users.

In February, members were urged to conserve electricity wherever possible, as hopes for a strike settlement appeared slim. The coal strike lasted for 110 days, ending on March 19, 1978.

As a response to rising rates, Ozark Electric offered some relief to members by approving levelized billing. This practice let members budget their electric bill by averaging the electric bill over the entire year. While bills would go up in low-use months, the extremely high bills in the winter and summer would go down.

Despite the energy crisis, rapid increases in wholesale rates, higher interest rates on loans, and major damage to the lines caused by storms, Ozark Electric survived the 1970s in excellent shape. This was pointed out by REA in a statistical report covering fifty-two items that demonstrated a borrower's ability to pay back its loans. The report concluded:

The presence of Table Rock Lake led to tremendous growth on the southern part of Ozark Electric's system beginning in the 1970s.

"Perhaps most significant is the fact that Ozark Electric Cooperative has been able to maintain its financial integrity with regard to mortgage and equity requirements in a period of drastic increases in labor, equipment, and, particularly, fuel costs. This helps to assure the members of Ozark Electric Cooperative of the availability of electric service at a lower cost than might otherwise be the case."

No one who was around in the 1970s will forget this time of great uncertainty and rising rates. No one liked passing on these rate increases any more than the members liked paying them. Yet members understood when given the information that rates were increasing through forces out of the board's and management's control.

"It was hard to explain on the phone, but if you went to a person's house and sat down you could explain it," Don Wood, Ozark Electric's assistant manager at the time, recalls of those difficult years. "Normally you could work through it and make them understand the reason rates were going up. But you couldn't make them like it."

Power lines were first placed underground during the 1970s, especially in subdivisions.

Ozark Electric's main office remains in Mount Vernon where it has been since the cooperative was organized.

Chapter 8
A Powerful Future

Good times would return for Ozark Electric Cooperative and its members. But first they would have to weather the farm crisis of the 1980s. In the 1970s, lowered trade barriers and increased purchases of farm products led to a dramatic increase in profits for farmers. In addition, land values kept going up. As a result, many farmers decided to increase their debt in order to enlarge their farming operations.

This agricultural boom didn't last long. In the 1970s, farm income was higher than the national average for the first time. But in the 1980s, the bubble burst. Income fell to 1960 levels as interest rates skyrocketed, doubling in less than a year. Farmland value dropped by nearly 60 percent in some parts of the Midwest between 1981 and 1985. And overproduction caused crop prices to fall. In a short time, more than one-third of the nation's farmers were in serious financial trouble.

Southwest Missouri was no exception, although the pattern was slightly different. Many farms that had been in the same family for generations went bankrupt. Families were forced off the farm or took to jobs in the city, farming part-time or not at all. All of the counties served by Ozark Electric lost farms, with Greene, Stone, and Christian Counties losing more than 10 percent. The fragile economies of small towns suffered too. Studies showed every farm that went bankrupt took with it three non-farm jobs.

Retired Director Bob Neal gained fame for portraying President Harry Truman. He's shown here visiting with the crowd at the Association of Missouri Electric Cooperative's annual meeting.

Ozark Electric Manager Pat Prewitt, who worked in engineering at the time, recalls actually hooking up farm services during this time period. He says a decade later, dairy farms that had produced one of the region's primary agricultural products began closing. "We came to the point where we were not seeing any new dairies," he recalls. "Dairies were good loads for Ozark Electric. At one time, Missouri was the number two dairy state and southwest Missouri was number one in Missouri. Later, poultry moved into the area."

Sparks/Record REA Edition, Sept. 1985, Page 6

Ozark Electric's employees

Management and employees of the Ozark Electric Cooperative in 1985 pose for a group photo. On the front row (left to right) are: DeLana Stokes, Jane Hill, Christine Robertson, Janet McElroy, Yvonne Johnson, Donna Mackey, Pat Killingsworth, Irene Tillman, Susan Langley and Betty Gayer. On the second row (from left) are: Don Wood, Jeff Langley, Buster Burns, Leon Wilson, Phil Callison, Roger Howell, Noel Wolf, Mike Meyers, Frank Smythia and Gary Davis. On the third row (from left) are: Marty Admire, Earl Wilson, Joe Fox, Lance Rycraft, Tom Shearrer, David Faucett, Don Russell, John Davis, Wes Caselman, Marlin Gray, C.W. Phillips and T.E. Sutton. On the fourth row (from left) are: Bill Johnston, Steve Swearingen, Richard Winn, Gene Messenger, Stacy Estes, Dennis Lackey, Bill McClain, B.J. Simkins, Mark Morris, Richard Neely and Jerry Mayberry. On the back row (from left) are: Wes Turner, Rick Frieze, Mike Killingsworth, Leo Losack, John Hall, Durrel Giest, Pat Prewitt, Brad Brown, John Brafford, Lee Coleman, Rick Sims and Denny Baker. Not pictured are Dewayne Austill and Thomas Lopez.

This photo from the Electric Sparks *shows Ozark Electric's employees in 1985.*

This dramatic decline in farming as a career led to a big change in the makeup of the cooperative's membership. Once a cooperative of farmers, Ozark Electric saw its membership increasingly composed more of retired people and those fleeing the city. Tempted by recreational opportunities at Table Rock Lake and Branson, people moved into Ozark's service area in droves. Another big influx came from the state of California, as residents of that state cashed in on the real estate boom and moved to Missouri where it was cheaper to live.

Between 1990 and 2000, Christian County's population increased 57.3 percent, while the population of Stone County increased 44.2 percent and Barry County increased by 23.5 percent. Between 2000 and 2010, Christian County was the state's fastest growing county, adding 23,137 people.

These changing demographics certainly changed the way members looked at their electric cooperative. With fewer people on the lines who remembered the dark days before electricity came to rural areas, Ozark Electric has had to work hard to keep its status as more than just a supplier of electricity.

"What's happening today is unfortunate. We are losing the people who remember when the lights came on," Pat says. "These people have always been the most dedicated to the cooperative idea. That generation is quickly disappearing. There are still quite a few that know what a cooperative is. But to many, we are just another utility. That is a challenge for us."

Don Wood recalls the story of an elderly woman who lived in the Cape Fair area and was an early member of the cooperative. One day her power went out. "She wrote a letter," the former assistant manager and lineman said. "In two to three days, that letter got to the office. 'I'm just letting you know I'm out of power.' Someone went down and fixed it. Today, five minutes is too long to be without power." That's one of the reasons Ozark Electric built two branch offices, the one in Cape Fair and the other, called the James River office, in Nixa.

In 1986, Jerry Mayberry became Ozark Electric's manager. The new manager's first job when he began working for Ozark Electric in June 1975 was to balance the books on the primitive computers then in use. That early system used two cassette tapes, one with a master file and the other a data file where changes were made. This information was then transferred over phone lines to Central Area Data Processing Cooperative based in St. Peters, Missouri.

Another issue Ozark Electric had to deal with was a large wave of retirements, especially in its outside workforce. "We were using some of the kids out of Manhattan, Kansas," Jerry Mayberry says of the young linemen Ozark hired out of tech school. "We had some come in and get their journeymanship and then go back home. We thought, 'why don't we send local kids to school, that way when they get their schooling they will come back home and stay here.'"

Ozark Electric Manager of Operations Earl Wilson joined the board at Linn State Technical College in Linn, Missouri, and in 2001 Ozark began offering two scholarships a year for the college's lineman training program to local people interested in becoming a lineman. Those accepted into the training program were hired by the cooperative when they graduated, provided there was an opening.

Deregulation became the buzzword in the electric utility industry in the 1990s. The idea touted by companies such as Enron was that by giving customers choice in who supplied their power, rates would go down. What they left unspoken was who would take care of customers no one wanted, and who would pay for upgrades to the electricity grid because those investments might never be paid back if the consumers it was built for left the system.

Several states passed laws making this a reality. Missouri's legislators, however, wisely decided to move slowly on the proposal. In every state that passed deregulation laws, rates went up and service suffered. This was one issue where Missouri's elected officials listened to the state's electric cooperatives and did not rush to pass deregulation measures.

Many electric cooperatives moved to diversify their businesses during those uncertain times, adding propane, satellite TV, or other related services to their electric business. But Ozark Electric's board stuck to its core business model—providing electricity—and did not get into these risky ventures.

Ozark did get into the fiber optic business, however. This move came more from a need to communicate with substations than to deal with the problem of potential deregulation of the industry. This business opportunity came when KAMO began installing fiber optic lines to remotely control its substations. KAMO offered to lease two pairs of fiber to Ozark from its backbone.

Jerry Mayberry speaks to the annual meeting crowd under the big top while the board backs him up.

Ozark Electric needed communication with its substations in order to utilize a new technology known as "AMR," or automated meter reading. Ozark Electric assisted Springfield City Utilities with the construction of fiber to serve the 911 call center located in Republic.

While deregulation was not the disaster it could have been, another potential issue loomed as the calendar rolled over to the year 2000. When the first computers came on line, they didn't have much memory. To save space, programmers shortened the year to the last two numbers. This would come back to haunt them as the industry began to worry that computers would not be able to distinguish between 1900 and 2000.

As the clock rolled past midnight on December 31, 1999, most of Ozark's staff was at the office holding a watch party just in case. There were no issues.

In 2006, Ozark Electric paid tribute to a number of pioneers who were instrumental in getting the cooperative off the ground or guided it through its greatest challenges. On July 18, 2006, the Verona substation was dedicated to Raymond Phariss, director from 1948 to 1972. Raymond attended the dedication, despite the fact he was about to turn 102 years old at the time. That same day, the Midway substation was dedicated to James C. Cox, who served as a director from 1958 to 2003. James's dad, W. C. Cox, was one of the twelve original founders of the cooperative.

The next day, July 19, 2006, the Lockwood substation was dedicated to William Arthur Carrier, who was a director from 1967 to 2006. Arthur served as board president and was a member of the KAMO Board of Directors, in addition to serving as KAMO's representative on the Associated Electric Board of Directors. The Greenfield substation was dedicated to Don A. Carlock, a director who served from 1975 to 2004. The Cave Springs substation was dedicated to Edward B. Staley, director from 1939 to 1977. He also served as board president.

In addition, two offices were dedicated. On August 25, 2006, the James River district office at Nixa was dedicated to Robert A. Neal, director from 1966 to 2006. In 2007, the Cape Fair district office was dedicated to Doug Stone, who was a director from 1967 to 2007.

In 2003, work began on installing automated meters across the system. For the first time since the early days, members would not have to read their own meters. Instead, the meters read

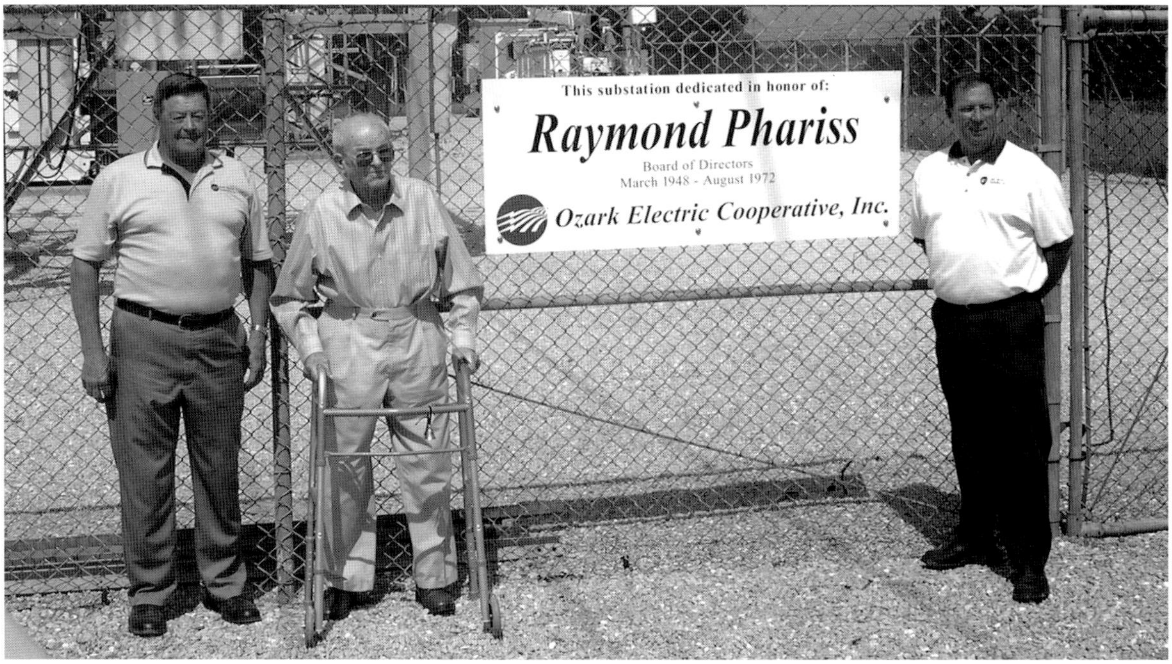

The Verona substation was dedicated to Raymond Phariss (center) in honor of his years of service on the Ozark Electric Cooperative Board. He served from 1946 to 1972 and was instrumental in signing up many early members. Also shown are Board Director Jimmie Crouch (left) and then General Manager Jerry Mayberry (right).

themselves and send the data back through the power lines to computers at the office. The last of these meters were installed in December 2006, ahead of schedule.

One month later, the new meters paid huge dividends when Ozark and other southwest Missouri electric cooperatives were struck by the worst ice storm in the history of the program. (See "The Big Ones" on pages 108-111) By sending a signal to the meters, called "pinging," employees could quickly determine who had power and who was without. The meters also continue to help with high bill complaints, allowing members to find patterns in their power consumption that could lead to better energy efficiency.

Recovering from the ice storm was a tremendous strain on the employees and members alike. A few months after spring 2007 finally arrived, Manager Jerry Mayberry decided to retire. "That took more out of me than I wanted," he said. "I had done this way too many times and I certainly had the years of service." His replacement was another veteran employee, Pat Prewitt, who came to Ozark Electric as staking engineer in August 1982. A graduate of Southwest Missouri State University in Springfield (now Missouri State), Pat first worked in a similar position at Ozarks Electric Cooperative in Fayetteville, Arkansas.

His arrival coincided with the rapid growth Ozark Electric was experiencing in the early 1980s. He remembers the cooperative making close to one thousand new connects per year. "We were picking up a lot of that Springfield growth," he says. "That's the area I worked."

In 1986, Pat moved to Ozark's James River office where he was district engineer and district manager. Later he was promoted to assistant manager and moved back to the Mount Vernon office.

The James River office located in Nixa opened in September 1987. It serves the many people who moved out of Springfield and settled on Ozark Electric's lines.

Bob Neal (center) reads the plaque dedicating the James River office to him. Neal, who served for many years as a director, was well respected in the Nixa area. He served as a director from 1966 to 2006. Then General Manager Jerry Mayberry (right) unveiled the plaque.

Ozark Electric's early linemen would never have dreamed lines would one day be worked from helicopters. This brave lineworker is installing marker balls that warn low-flying aircraft of power lines.

N1097J

This photo shows the tremendous amount of ice that collected on lines during the 2007 ice storm. The weight broke poles by the thousands.

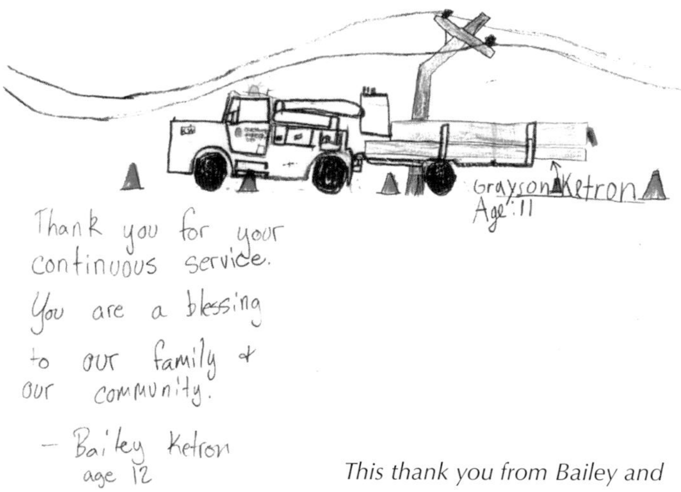

This thank you from Bailey and Grayson Ketron was one of many the cooperative received in the wake of the devastating ice storm in 2007.

One thing Pat Prewitt has worked hard on is to improve relations with neighboring utilities, including Springfield City Utilities, Empire Electric, Mount Vernon, and Monett. "When I first came to work here, we were fierce competitors," he says. "We still compete for loads, but now there's a new age of cooperation. Now we have a relationship where we can discuss any utility issues. We are working together to provide what the public needs."

In his five years as manager, Pat has seen more pressure on rates as Congress began debating the climate change issue. This was a special concern for electric utilities in Missouri and other midwestern states because 80 percent of the electricity used is generated by coal.

Also cause for concern was the 2 percent per year growth in electricity use, with many consumers adding load from a variety of devices on the market. Cell phones, iPods, big-screen high-definition TVs…they all had one thing in common, a cord to plug them into the grid.

At the same time Associated Electric needed to plan for the next generation of baseload power plants, legislative uncertainty made it difficult to choose what type of plant would be the best for the future. In fact, the Associated board canceled plans for a coal-fired power plant to be located in Norborne.

Several steps were taken to deal with this issue. First, Associated Electric continued to diversify its generation facilities, purchasing the entire output from four wind farms located in northwest Missouri. Because of the intermittent nature of wind power—it blows the least when demand for electricity is the highest—these projects cannot be considered baseload generation.

This dramatic photo taken by Ozark Electric Chief Financial Officer Tony Forquer ran in publications and on websites across the nation. It best illustrates the incredible damage wrought by the 2007 ice storm.

Ozark Electric opened a district office in Cape Fair in September 1993 to better serve the growing membership in the Table Rock Lake area.

General Manager Pat Prewitt and Manager of Engineering Ken Raming oversee a conductor change at a lake crossing.

Associated Electric bought an unfinished natural gas plant in Dell, Arkansas, and finished building it in 2009. In addition, a mothballed gas plant, often referred to as a "plant in a box," was purchased for pennies on the dollar. It was rechristened Chouteau II and assembled next to an existing plant in Oklahoma.

These and smaller gas-fired peaking units offered enough power to get the electric cooperatives to the 2023 timeframe before the need for more baseload power will be required. But because power plants take many years to build, the Associated board continues to grapple with a long-range plan for future watts, and it seems clear they will have fewer options for the future.

Nuclear power is being considered because it doesn't release carbon dioxide like coal and natural gas plants. Nuclear power plants are expensive to build, but once on line, the plants are extremely reliable and provide low-cost electricity.

"Future power costs, due to environmental regulations, are by far the number one issue for electric cooperatives today," Pat says. "It's consistently discussed whenever our board meets. Today, 53 to 55 percent of every dollar goes to wholesale power costs."

He believes nuclear power will have to play a role in future power generation. "We need to move forward with that," he says. "Nuclear power has provided some of the cheapest energy in the long run and it's clean. They don't like to call it green power, but I do. The only thing coming out of the stacks is water vapor."

Ozark Electric felt the loss of another faithful employee when Member Services Department Manager Denny Baker retired on August 1, 2009. For many years, Denny was the face and voice

Manager Pat Prewitt takes his place at the podium during Ozark Electric's 2011 annual meeting. He was hired as a staking engineer in August 1982 and became manager in 2007.

From 1984 until his retirement in 2009, Denny Baker was the face and voice of Ozark Electric. He is shown here with 2005 Youth Tour delegates Valerie Melton, Mount Vernon; Ben Poirot, Greenfield; and Stefan Bellm, Nixa.

of Ozark Electric. Whether he was visiting members' homes to help them lower their electricity use, working with students in FFA and Youth Tour programs, or planning twenty-five years worth of successful annual meetings, he always understood and implemented the cooperative's philosophy of knowing he worked for the membership and the member came first. He also organized many employee functions, including the annual picnics and Christmas banquets and was known as the "class clown" by fellow employees.

Today, Ozark Electric Cooperative remains one of the state's strongest and largest electric cooperatives. With 31,000 meters, it is the state's seventh largest. It's the eighth largest in miles of line, with 4,763.

Those pioneers such as W. L. Boucher and Irby Murray would be amazed at the powerful equipment and the ninety well-trained employees who keep the cooperative financially sound and the power flowing. They would be proud of the miniscule 5 percent line loss, a number it would be difficult to improve upon. They would be envious of the 6.5 members-per-mile density and would be stunned at the average of 1,105 kilowatt-hours the typical member uses each month.

But one thing they would certainly recognize is the cooperative way of doing business that has continued for seventy-five years. No one can accurately predict the future for Ozark Electric Cooperative. But one thing is certain. As long as it continues to operate solely to provide service to its members, as long as it is governed by its own members through an elected board of directors, and as long as it maintains the ideal that all people deserve access to electricity at the lowest possible cost, Ozark Electric has a bright and powerful future.

Employees of Ozark Electric turn out in force for the American Cancer Society's Relay for Life. This crew was part of the 2001 relay.

The equipment has changed dramatically, but linemen today remain dedicated to keeping the power flowing, no matter how hard the job may be. Shown here are, from left, Roger Lester, Billy Bellah, Win Osteloh, and John Banning.

Ozark Electric's linemen pose with their trucks in this 2011 photo. The equipment they use would amaze early linemen.

The ice storm that struck in 1972 was considered the worst natural disaster to hit Ozark Electric until the far worse 2007 storm.

The Big Ones

Since the beginning of Ozark Electric Cooperative, weather has plagued the mightiest efforts of its crews to keep power flowing. From a few dozen outages to several thousand, ice storms, tornadoes, lightning, and floods have caused outages on the cooperative's lines.

One of the earliest disasters to strike the co-op came in 1949, when heavy ice blanketed much of the state. In those days members could be a lot more patient when waiting for service, since few had more than a few lights powered by electricity and many had retained their coal oil lanterns. It wasn't unusual then for members to send postcards to their cooperative asking them to restore power when they were out that way.

Sue Childress Compton remembers those early ice storms knocking out phone service as well as the electricity. "They would take me to the telephone office and I would take trouble calls there. That happened to me at least two times."

The 1949 ice storm was the first test of the statewide "Procedure for Emergency Assistance" put into place at the Association of Missouri Electric Cooperatives in Jefferson City at a meeting held in December 1948. Under that plan, any system needing help to restore power could call on electric co-ops located out of harm's way.

In the ensuing years, Ozark Electric had been a beneficiary of this emergency assistance and also one of the systems always willing to lend a hand. In fact, crews have traveled as far away as the Gulf Coast to help restore power for electric co-ops in Alabama and Louisiana devastated by Hurricanes Ivan, Katrina, and Rita.

As we worked through storm issues, way
nore than I care to remember, the best thing that ever happened was Statewide putting together he assistance," says former Ozark Manager Jerry Mayberry. "All you had to do is make that one phone call and assistance from sister cooperatives was on its way."

Up until recently, all major outages on Ozark Electric's lines were put into perspective by the ice storm of 1972. Winter that year got an early start when a heavy snow blanketed the area on November 18–20. The weight of the snow caused trees to sag nto lines, toppling poles and snapping wires.

All thoughts of that damage were forgotten a nonth later when ice began to fall on December 1. For three days, crews lost ground on the storm, with outages rising too fast to count. Poles toppled ike dominoes, and outside crews from as far away as Cameron in north Missouri rushed in to assist the beleaguered Ozark Electric and KAMO linemen. Some ninety-six seasoned linemen came to their aid along with another fifteen who helped with radio communications.

Four electrical contractors were called in for service. White River Valley, Barry Electric, and Northeast Electric in Oklahoma and the City of Crane sent crews. Other cooperatives sent supplies, ncluding a trailer load of poles from Southwest Electric that helped replace the 431 that were lost.

The National Guard brought portable generators o help farmers—now dependent on electricity—milk their cows. The National Guard also used its helicopters to patrol KAMO Power's transmission ines, while the highway patrol and county sheriff departments helped with communications and highway crossings.

That storm caused $325,000 in damage. Not long after, the cooperative was rocked by a tornado at Lockwood and eighteen inches of rain fell around Boaz, washing out several poles.

But in 2007, a new standard for devastation was recorded when an ice storm following the identical

Justin Bass and Darren Deal were two Ozark Electric linemen who traveled south to assist electric cooperatives affected by Hurricane Ivan.

Beautiful, but deadly. The ice storm of 2007 left trees and poles alike blanketed with a heavy layer of ice. Much of Ozark Electric's service area returned to the dark days as lines crashed down in waves.

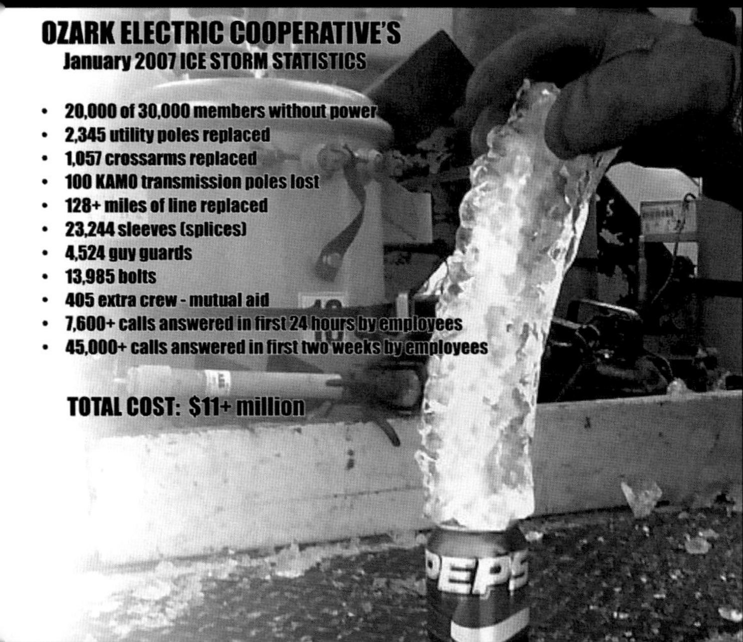

This soda can illustrates how thick the ice blanketing Ozark Electric lines was in the wake of the 2007 storm. Photo courtesy of Ozark Electric Lineman Chip Beerbower.

path of the 1949 storm struck Ozark Electric with a vengeance. The first outage reports began coming in Friday night, January 12. Wave after wave of freezing precipitation kept coming. On Sunday night, January 14, Assistant Manager Pat Prewitt and Member Services Manager Denny Baker went out to survey the damage. In one area alone, just north of the Mount Vernon substation, they counted approximately seventy-eight poles in a five-mile stretch laid flat by the storm.

The 2007 storm caused lengthy outages for 120,000 electric cooperative members statewide. Ozark Electric was one of the hardest hit, with miles of lines completely toppled and thousands of poles snapped off like twigs. By Monday morning, more than 20,000 of its 30,000 members were without power.

Traveling down Interstate 44 west of Springfield, the area was completely dark in the wake of the storm. Shattered trees reminded World War II veterans of bombed-out parts of Nazi Germany. In one instance, a two-mile section of Ozark's three-phase line collapsed. Every pole along the line lay on the ground. Similar damage was reported in much of Ozark's nine-county area. All told, the co-op replaced more than two thousand poles in fourteen days.

"When a wire breaks, all the weight on the other side pulls real hard on that pole. When the pole breaks, it puts weight on the next pole," explained John Davis, who was then Ozark's operations manager, in a story in *Rural Missouri* magazine. "You'll lose 10, 12 poles before it gets enough weight off. You'd get one span up and the next span would go down. You go down and put it back up and watch it go down in another spot."

At its darkest hour, Ozark Electric would receive fresh crews from the southern electric cooperatives they helped during the hurricanes. Stunned by the bitter cold, the southerners purchased warmer clothes and quickly turned the tide against the damage. One Mississippi lineman working the outage was heard to remark, "I'd rather work 15 Hurricane Katrinas than one day in this ice!" More than 450 linemen, including workers from all over

Missouri plus Louisiana, Mississippi, Alabama, Iowa, and Kentucky, assisted Ozark Electric.

It would take two weeks for all lines to be once again energized, and more than a year before all of the repairs to the patched-up system could be completed. What had begun as the cooperative's darkest moment had turned into its finest hour, thanks to the help of some friends. Ozark Electric's crews would again return the favor when an even more devastating ice storm hit south-central and southeast Missouri, this time sparing Ozark Electric by a few miles.

Outages have been caused by a number of factors. The most unusual occurred in 1974, when crews working for Springfield City Utilities set off explosives under one of KAMO's 69,000-volt transmission lines. The errant blast knocked out half of Ozark Electric's system for as long as an hour until repairs could be made.

While ice storms have been the worst culprits for power outages, tornadoes have done significant damage. One of the worst took place May 4, 2003. The twister demolished parts of Battlefield, Pierce City, and Stockton and caused widespread damage to Ozark Electric's lines. Fortunately, most Ozark members were outside the path of the EF5 tornado that devastated Joplin on May 22, 2011.

Whatever the cause and level of destruction, Ozark Electric's crews stand ready to restore power no matter the cost. Its members always seem to rise to the occasion as well, and many do their part to help the cooperative's employees make it through the difficult times by bringing hot drinks, snacks, and words of encouragement in the form of cards, letters, and drawings from the kids.

One comment from the 2007 ice storm, sent in by Ozark members J. F. and Avis Bradshaw, sums up the feeling of pride members have in their cooperative: "And then on that glorious day when the power came back on, we realized that real heroes ride in white trucks and wear hard hats and have hearts of gold to be of service and care for their fellow men."

This image was used in a variety of ways to tell the story of the great ice storm of 2007.

Battlefield was one city in the path of an outbreak of tornadoes on May 4, 2003. The twister caused extensive damage there and also in Pierce City and Stockton.

Ozark Electric Cooperative Timeline

1937 March 16, 1937: First meeting of those interested in forming an electric cooperative held in Mount Vernon. Approximately fifty attended.

October 21, 1937: First meeting held to officially organize the cooperative. Officers elected were Arley Brown, president; Irby Murray, vice president; and Justin Carlock, secretary/treasurer.

Original incorporators were Arley Brown, Aurora; Irby Murray, Ash Grove; Justin Carlock, Everton; Forest Speight, Everton; Roy Sifferman, Republic; T. O. Holman, Stotts City; W. C. Cox, Aurora; J. B. Williams, La Russell; and C. A. Sifferman, Mount Vernon. W. L. Boucher was named project manager.

The remains of this early Ozark Electric truck can be found near Crane.

November 1, 1937: Sverdrup & Parcel of St. Louis hired as project engineer.

1938 February 7, 1938: G. Seay hired as inspector. Miss Louvina Watterson hired as stenographer and bookkeeper.

March 10, 1938: Ozark Electric joins the Missouri State Rural Electrification Association, with W. L. Boucher serving as its representative.

April 21, 1938: Williams Brothers and Haas Construction of Chicago and Tulsa selected to build first line.

May 5, 1938: Hubert Jones of Everton hired as Ozark Electric's first lineman at $100 a month.

June 18, 1938: Board approves $5,000 loan from REA to finance wiring and plumbing for members.

June 28, 1938: Ozark Electric buys its first truck, an International with a McCabe-Powers utility bed, for $275 from Sifferman Brothers Motor & Implement in Mount Vernon.

August 12, 1938: Ozark Electric rents office space in the Ray Richardson Building across from the post office in Mount Vernon.

September 23, 1938: Power flows for the first time on the Verona-Red Oak section of line, the first REA line in southwest Missouri. Lawrence County Extension Agent John Woodward had the honor of throwing the first switch.

October 20, 1938: An eleven-pole line extension built to serve the Missouri State Sanatorium farms.

1939 March 8, 1939: Irby Murray becomes board president after Arley Brown resigns. Loan to build Section B approved by REA.

June 1939: A total of $1,136.64 transferred to New-Mac Electric from memberships made by 384 prospective members who are now part of this new REA project. Director A. N. Walker resigns to join the New-Mac board.

The longest continuous connection to the cooperative is the Dadeville United Methodist Church, which received power on October 1, 1938, and has continued as a member of Ozark Electric ever since.

July 31, 1939: Members required to read their own meters.

October 10, 1939: A total of $1,545 transferred to Southwest Electric for prospective members who signed on to Section D and are now part of this new cooperative. Bids opened for Section C line construction.

1940 January 2, 1940: Ozark Electric now has 1,700 members.

January 22, 1940: W. L. Boucher resigns as project manager. Board names Assistant Project Manager S. E. Roberts as his replacement.

January 30, 1940: Minimum bill reduced by 50¢ to $5.50.

June 1940: Cooperative buys electric roasters to use in cooking demonstrations.

August 8, 1940: Money paid by prospective members to the north transferred to Sac Osage Electric.

September 9, 1940: White River Valley Electric receives $287 transferred from former Ozark members who are now part of that project.

1941 May 2, 1941: Board approves $100 membership to join KAMO Power.

June 2, 1941: Co-op asks to borrow $35,000 from REA to help finance the installation of indoor plumbing for three hundred members. Loan later increased to $40,000.

August 2, 1941: With a large balance in the general operating fund, the cooperative makes a $15,000 advance payment to REA.

October 2, 1941: Mr. and Mrs. Hal Childress donate three acres of land to the cooperative for the new office building.

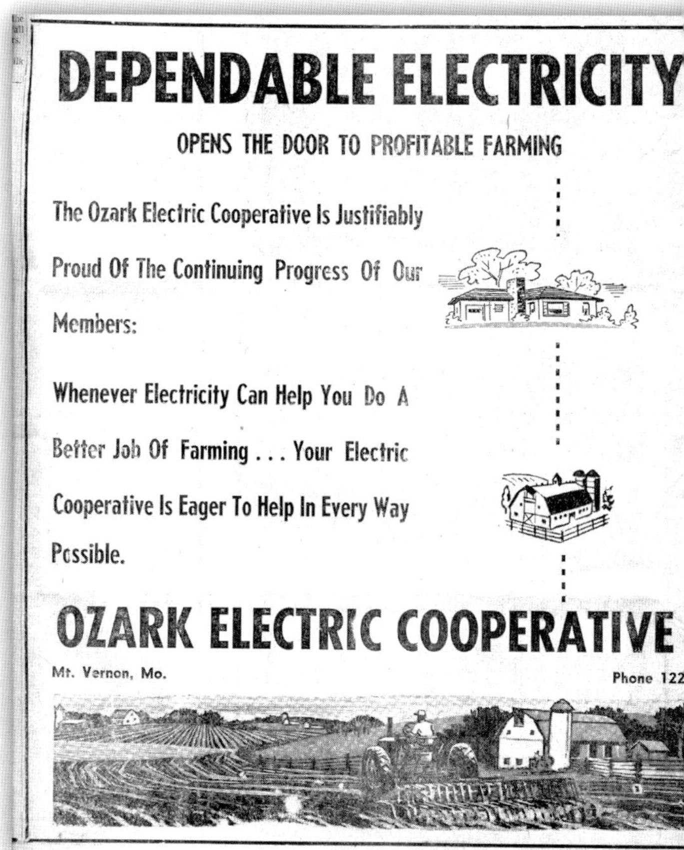

Ads such as this one printed in 1955 showed the benefits of electricity on the farm. At this time the cooperative was working to build load.

November 1941: Hot plates given to every school and church that made holiday decorations using electric cooperative current.

1942 March 2, 1942: Cooperative has 2,562 members.

September 2, 1942: Ozark Electric joins the safety and training program of the Missouri State Rural Electrification Association. The cooperative also becomes a member of the National Rural Electric Cooperative Association. Diggins, Fordland, Rogersville, and Seymour added to Ozark Electric's service area.

October 1942: Co-op moves into new office north of Mount Vernon.

October 26, 1942: Special meeting held to convert corporation into a cooperative, non-profit, membership corporation pursuant to Section 16 of the Rural Electric Cooperative Act.

1943 January 13, 1943: Co-op buys $7,000 in Victory Bonds.

March 27, 1943: Annual meeting talk by Lawrence County Extension Agent Ray Graham, "What this cooperative can do toward production for the war." Co-op has 3,236 members.

December 8, 1943: Manager S. E. Roberts given a one-year leave of absence to take over as manager of Sho-Me Power. Later he was released as manager and A. Claypool Daughtery of Lamar was hired as his replacement effective March 15, 1944.

1944 April 12, 1944: Board guarantees a loan from REA to establish refrigeration cooperatives in Mount Vernon and Marionville. This was intended to help provide more food for local consumption and to alleviate the labor shortage caused by the war.

1945 February 14, 1945: George Robbins resigns from the board in order to accept a position as manager of New-Mac's Cassville office.

June 6, 1945: Members of Ozark Electric now in the service area of New-Mac Electric in Barry County were released to that cooperative.

August 8, 1945: Ozark agrees to lease from Sho-Me all lines in the service area. Also agrees to take over Sho-Me loans in the amount of $90,000 to build lines in Webster, Christian, and Douglas Counties to serve 236 members.

1947 December 10, 1947: Ozark agrees to join other co-ops in paying for KAMO's development costs prior to getting power from the generation and transmission cooperative.

1948 April 14, 1948: Ozark agrees to subscribe to the *Rural Electric Missourian* for 55¢ per year for eleven copies for each member.

May 12, 1948: Ozark Electric applies to REA for funds to install two-way radios in line trucks.

Ozark Electric had a fleet of these Chevy trucks in the 1950s.

September 1948: Doris Poindexter begins duties as Ozark's home advisor in the power-use department. A model kitchen designed for cooking demonstrations was built at the American Legion Hall in Mount Vernon for her use.

A large banner welcomes members to the 2004 annual meeting.

1949 January 12, 1949: Board agrees to take part in state emergency assistance plan.

1953 September 1953: Claypool Daugherty resigns as manager. Marvin Carver takes over as acting manager.

December 1953: S. E. Roberts returns as manager after ten years managing Sho-Me Power.

1954 April 8, 1954: Monett substation is the first energized from KAMO's transmission lines.

1955 July 20, 1955: The Miller substation, the last substation served by Empire, was transferred to KAMO Power. Ozark Electric is now receiving all of its wholesale power from the cooperative it helped create.

1964 June 1964: Chester Swearingen becomes the new manager after the death of S. E. Roberts.

1965 May 1965: Charlotte Ann Burns and Greg Lynn Curl selected as Ozark Electric's first delegates to the Rural Electric Youth Tour to Washington, D.C.

1972 December 11, 1972: Worst ice storm in Ozark Electric's history strikes.

1974 January 1974: Work begins on a new office building on the site of the old one. Staff move into a trailer while construction takes place.

July 15, 1974: Tommy Sutton replaces Chester Swearingen as manager.

April 1974: Computerized billing begins.

1985 October 1985: Don Wood becomes acting manager after Tommy Sutton resigns.

1986 June 1986: Jerry Mayberry becomes the new manager.

1987 March 1987: Ozark Electric announces the *Electric Sparks* newsletter would become part of the *Rural Missouri* magazine. Former Ozark Electric Assistant Manager Don Wood becomes the magazine's new editor.

September 1987: James River office at Nixa opens.

1993 September 1993: Cape Fair office opens.

2006 December 2006: Installation of automated meters completed ahead of schedule, ending the need for members to take meter readings.

2007 January 12, 2007: An ice storm far worse than the 1972 storm strikes Ozark Electric, forcing more than 20,000 outages and two weeks of repairs.

August 2007: Pat Prewitt becomes manager after Jerry Mayberry retires.

2012 October 21, 2012: Ozark Electric celebrates its seventy-fifth anniversary.

Willie Wiredhand celebrates Ozark Electric's seventy-fifth anniversary in this window at the headquarters building in Mount Vernon.

Board of Directors, clockwise from left:
Director Ron Crighton, Director James Stine, Director Dwain Hughes, CFO Tony Forquer, Attorney Rod Widger, Board Vice President Gary Duvall, Board President David Trogdon, Board Secretary/Treasurer Steve Stone, Board Secretary DeLana Stokes, Director O. D. Cope, Director Jimmie Crouch, Director Earl Carpenter, Cooperative General Manager Patrick L. Prewitt.

Operations and Engineering:
Front row: Phillip Hood, Jacob Bradley, Nick Shiveley, Vince Robison, Drew Meier, Gene Messenger, Mike Clevenger, Justin Bass, Dustin Bray, Craig Rice, Justin Rock, Jason Wilson, Billy Bellah, Kelly Essary, Brandon Cole, Tobey Bennett, Phillip Elbert, Robin Childs, Jeff Smith, Justin Fitzpatrick, Roger Lester, Brian Cutbirth, and Daniel Smith, Back Row: Clayton Coble, Ed Harter, Dan Lohkamp, Curt Straw, Dustin Russell, Matt Ogden, Chris Martin, Mike Baker, Tom Shearrer, Ray Patrick, Rick Sims, Brad Brown, Jimmie Moore, Jeff Strain, Dean Branstetter, John Banning, Jacob Jones, J .R. Bowen, Win Osterloh, Earl Wilson (on truck), Neal Moore, Kevin Burnett, Chad Conner, Terry Winingham, Ray Rogers, Rick Palmquist (in truck), Steve Swearingen, Don Jester, and Mike Woodrum.

Altec

Cape Fair District Office Employees:
Row 1: Jodie Essary, Robin Dunagan;
Row 2: Kelly Essary, Craig Rice, Mike Clevenger, Dustin Bray, Brian Cutbirth;
Row 3: John Banning, Nick Shiveley, Jeff Sullivan, Dustin Russell, Kevin Burnett, Curt Straw.

Cape Fair District Office Employees

James River District Office at Nixa Employees:
Row 1: Jamie Matlock, Ginger McLemore, Christine Robertson, Kathy Gann-Willig, Jane Brewer, Connie Gilmore;
Row 2: Stacy Estes, Chris Martin, Jacob Bradley, Justin Rock, Andy Groesbeck, Jason Wilson, Brandon Cole;

Row 3: Dan Lohkamp, J. R. Bowen, Rick Sims, Tobey Bennett, Jeff Smith, Justin Fitzpatrick;
Row 4: Jeff Strain, Chad Conner, Matt Ogden, Terry Winingham, Steve Swearingen, Engineering Manager Ken Raming.

Mount Vernon Headquarters Employees:
Row 1: General Manager Pat Prewitt, Janet McElroy, Yvonne Johnson, Holly Hodgson, Cindy Freitag, Shirley Barnes, Summer Collier, Linda Stander;
Row 2: Wes Dyson, Robin Childs, Kem Hall, Dora Bell, Human Resources Manager DeLana Stokes, Suzanne Stockton, Donna Cole, Patty Killingsworth, Barbara Coy;
Row 3: Phillip Hood, Don Jester, Drew Meier, Billy Bellah, Daniel Smith, Neal Moore, Phillip Elbert, Vince Robison, Finance Manager Tony Forquer, Operations Manager Earl Wilson;

Row 4: Consumer Services Manager Patrick Oehlschlager, Kevin Fulks, John Carey, Jim Barb, Mike Baker, Roger Lester, Jimmie Moore, Jacob Jones, Win Osterloh, Mike Woodrum, Ray Patrick, Rick Palmquist, Kent Rust;
Row 5: Clayton Coble, Dean Branstetter, Ed Harter, Ray Rogers, Brad Brown, Bruce Roseman, Darren Deal, Justin Bass, Gene Messenger.
(Not pictured: Chip Beerbower, Tom Shearrer, Richard Winn)

About the Author

Jim McCarty, fifty, of Taos, Missouri, is the editor of *Rural Missouri* magazine, the official publication of the Association of Missouri Electric Cooperatives. He joined the staff in November of 1985 as assistant editor and quickly moved up to managing editor, then editor in 1987. He oversees a staff of six who have won hundreds of awards for photography, writing, and publications. Jim was the writer of the year for the Cooperative Communicators Association in 1994 and in 1995 won that group's Michael Graznak Young Communicators Award.

He is a 1984 honor graduate of the University of Missouri with a degree in photojournalism. He went to high school in St. Clair, Missouri, graduating with honors in 1980 and is a 1982 graduate of East Central College in Union, Missouri. His journalism career began at the *Daily American Republic* newspaper in Poplar Bluff, Missouri, where he was assistant sports editor and photographer.

Jim helped found the National Country Market advertising cooperative and is a past member of that organization's board of directors. He is past president of the National Electric Cooperative Statewide Editor's Association and past president of the Missouri Association of Publications, another organization he helped found.

He is the author of five books on the history of electric cooperatives: *Let There Be Light* (Callaway Electric), *Bigger Isn't Always Better* (Howard Electric), *Rural Wonder* (Farmers' Electric), *From Darkness to Daylight* (Tri-County, Minnesota), and *Small but Mighty* (Missouri Rural Electric). In 2009 he documented the ice storm that hit southern Missouri and wrote another book on the recovery effort called *Winter's Wrath*. He also was instrumental in producing "Use or Usage: A Rural Electric Guide to Style" for the Statewide Editors Association.

In 1995 he won the Joe Humble Newsletter Editor Award presented by the Artist-Blacksmith's Association of North America (ABANA) and in 1996 he shared the Blacksmiths Association of Missouri's (BAM) Bob Patrick Founder's Award with his brother, Pat. In 1998 he was honored with BAM's lifetime membership award.

From 1997 to 2000 he was editor of *The Anvils Ring* and the *Hammer's Blow*, two quarterly blacksmithing magazines published by ABANA.

An avid runner, he has finished six marathons and three ultramarathons. In his spare time he operates the Father Helias Forge where he does blacksmithing and woodworking.

Jim is married to Gretchen and has three children, James, Daniel, and Emily.

Altec